Contaminated Sediments in Ports and Waterways

Cleanup Strategies and Technologies

Committee on Contaminated Marine Sediments

Marine Board
Commission on Engineering and Technical Systems
National Research Council

NATIONAL ACADEMY PRESS
Washington, D.C. 1997

NATIONAL ACADEMY PRESS • 2101 Constitution Avenue, N.W. • Washington, DC 20418

NOTICE: The project that is the subject of this report was approved by the Governing Board of the National Research Council, whose members are drawn from the councils of the National Academy of Sciences, the National Academy of Engineering, and the Institute of Medicine. The members of the panel responsible for the report were chosen for their special competencies and with regard for appropriate balance.

This report has been reviewed by a group other than the authors according to procedures approved by a Report Review Committee consisting of members of the National Academy of Sciences, the National Academy of Engineering, and the Institute of Medicine.

The program described in this report is supported by Cooperative Agreement No. DTMA91-94-G-00003 between the Maritime Administration of the U.S. Department of Transportation and the National Academy of Sciences. Any opinions, findings, conclusions, or recommendations expressed in this publication are those of the author(s) and do not necessarily reflect the view of the organizations or agencies that provided support for the project.

Library of Congress Cataloging-in-Publication Data

Contaminated sediments in ports and waterways : clean-up strategies and technologies / Committee on Contaminated Marine Sediments, Marine Board, Commission on Engineering and Technical Systems, National Research Council.
    p. cm.
  Includes bibliographical references and index.
  ISBN 0-309-05493-1(alk. paper)
  1. Contaminated sediments—Management. 2. Marine sediments.
3. Harbors. 4. Waterways. I. National Research Council (U.S.).
Committee on Contaminated Marine Sediments.
TD878.C665 1997
363.739'4—dc21   96-52050
                                                                                                     CIP

Copyright © 1997 by the National Academy of Sciences. All rights reserved.

Printed in the United States of America

COMMITTEE ON CONTAMINATED MARINE SEDIMENTS

HENRY J. BOKUNIEWICZ, *co-chair,* State University of New York at Stony Brook
KENNETH S. KAMLET, *co-chair,* Linowes and Blocher, Silver Spring, Maryland
W. FRANK BOHLEN, University of Connecticut, Groton
J. FREDERICK GRASSLE, Rutgers University, New Brunswick, New Jersey
DONALD F. HAYES, University of Utah, Salt Lake City
JAMES R. HUNT, University of California, Berkeley
DWAYNE G. LEE, Parsons Infrastructure and Technology Group, Pasadena, California
KENNETH E. MCCONNELL, University of Maryland, College Park
SPYROS P. PAVLOU, URS Greiner, Inc., Seattle, Washington
RICHARD K. PEDDICORD, EA Engineering, Science, and Technology, Hunt Valley, Maryland
PETER SHELLEY, Conservation Law Foundation, Inc., Boston, Massachusetts
RICHARD SOBEL, Clean Sites, Inc., Alexandria, Virginia
LOUIS J. THIBODEAUX, Louisiana State University, Baton Rouge
JAMES G. WENZEL, NAE, Marine Development Associates, Inc., Saratoga, California
LILY Y. YOUNG, Rutgers University, New Brunswick, New Jersey

Liaison Representatives

SABINE APITZ, Naval Command, Control, and Ocean Surveillance Center, San Diego, California
CHARLES C. CALHOUN, U.S. Army Engineer Waterways Experiment Station, Vicksburg, Mississippi
MILES CROOM, National Marine Fisheries Service, Silver Spring, Maryland
ROBERT ENGLER, U.S. Army Engineer Waterways Experiment Station, Vicksburg, Mississippi
KENNETH HOOD, Environmental Protection Agency, Washington, D.C.
EVIE KALKETENIDOU, Maritime Administration, Washington, D.C.
DANIEL LEUBECKER, Maritime Administration, Washington, D.C.
FRANK MANHEIM, U.S. Geological Survey, Woods Hole, Massachusetts
JANET MORTON, U.S. Geological Survey, Reston, Virginia
ANNA PALMISANO, Office of Naval Research, Arlington, Virginia
CARL SOBREMISANA, Maritime Administration, Washington, D.C.
MARK SPRENGER, Environmental Protection Agency, Edison, New Jersey
CRAIG VOGT, Environmental Protection Agency, Washington, D.C.
LARRY ZARAGOZA, Environmental Protection Agency, Washington, D.C.

Staff

JOSEPH L. ZELIBOR, JR., Study Director
KRISTIN CHURCHILL, Program Associate
DELPHINE D. GLAZE, Administrative Assistant (since June 1995)
RICKY PAYNE, Administrative Assistant
SHARON RUSSELL, Administrative Assistant
LAURA OST, Editor
WYETHA B. TURNEY, Production Assistant

MARINE BOARD

JAMES M. COLEMAN, NAE, *chair*, Louisiana State University, Baton Rouge
JERRY A. ASPLAND, *vice-chair*, California Maritime Academy, Vallejo
BERNHARD J. ABRAHAMSSON, University of Wisconsin, Superior
BROCK B. BERNSTEIN, EcoAnalysis, Ojai, California
LILLIAN C. BORRONE, NAE, Port Authority of New York and New Jersey
SARAH CHASIS, Natural Resources Defense Council, New York
CHRYSSOSTOMOS CHRYSSOSTOMIDIS, Massachusetts Institute of Technology, Cambridge
BILIANA CICIN-SAIN, University of Delaware, Newark
BILLY L. EDGE, Texas A&M University, College Station
JOHN W. FARRINGTON, Woods Hole Oceanographic Institution, Woods Hole, Massachusetts
MARTHA GRABOWSKI, LeMoyne College and Rensselaer Polytechnic Institute, Cazenovia, New York
JAMES D. MURFF, Exxon Production Research Company, Houston, Texas
M. ELISABETH PATÉ-CORNELL, NAE, Stanford University, Stanford, California
DONALD W. PRITCHARD, NAE, State University of New York at Stony Brook and Severna Park, Maryland
STEVEN T. SCALZO, Foss Maritime Company, Seattle, Washington
MALCOLM L. SPAULDING, University of Rhode Island, Narragansett
KARL K. TUREKIAN, NAS, Yale University, New Haven, Connecticut
ROD VULOVIC, Sea-Land Service, Charlotte, North Carolina
E.G. "SKIP" WARD, Shell Offshore, Houston, Texas

Staff

CHARLES A. BOOKMAN, Director
DONALD W. PERKINS, Associate Director
DORIS C. HOLMES, Staff Associate

The National Academy of Sciences is a private, nonprofit, self-perpetuating society of distinguished scholars engaged in scientific and engineering research, dedicated to the furtherance of science and technology and to their use for the general welfare. Upon the authority of the charter granted to it by the Congress in 1863, the Academy has a mandate that requires it to advise the federal government on scientific and technical matters. Dr. Bruce M. Alberts is president of the National Academy of Sciences.

The National Academy of Engineering was established in 1964, under the charter of the National Academy of Sciences, as a parallel organization of outstanding engineers. It is autonomous in its administration and in the selection of its members, sharing with the National Academy of Sciences the responsibility for advising the federal government. The National Academy of Engineering also sponsors engineering programs aimed at meeting national needs, encourages education and research, and recognizes the superior achievements of engineers. Dr. William A. Wulf is interim president of the National Academy of Engineering

The Institute of Medicine was established in 1970 by the National Academy of Sciences to secure the services of eminent members of appropriate professions in the examination of policy matters pertaining to the health of the public. The Institute acts under the responsibility given to the National Academy of Sciences by its congressional charter to be an adviser to the federal government and, upon its own initiative, to identify issues of medical care, research, and education. Dr. Kenneth I. Shine is president of the Institute of Medicine.

The National Research Council was organized by the National Academy of Sciences in 1916 to associate the broad community of science and technology with the Academy's purposes of furthering knowledge and advising the federal government. Functioning in accordance with general policies determined by the Academy, the Council has become the principal operating agency of both the National Academy of Sciences and the National Academy of Engineering in providing services to the government, the public, and the scientific and engineering communities. The Council is administered jointly by both Academies and the Institute of Medicine. Dr. Bruce M. Alberts and Dr. William A. Wulf are chairman and interim vice chairman, respectively, of the National Research Council.

Preface

BACKGROUND

Contaminated marine sediments threaten ecosystems, marine resources, and human health. Sediment contamination also can have major economic impacts when controversies over risks and costs of sediment management interfere with the regular and periodic need to dredge major ports. Approximately 95 percent (by weight), or 1.4 billion tons, of total U.S. trade passes through dredged ports (Maritime Administration, 1994).

In a previous report, the National Research Council (NRC) (1989) examined the extent and significance of contaminated marine sediments, the state of practice of cleanup and remediation, and management strategies. Although contamination was a serious concern at many sites, the report found that remedial action was rare. Several barriers to remediation were identified, including insufficient data for the comprehensive listing and prioritization of contaminated sites, the lack of widely accepted techniques for identifying and assessing contamination in marine sediments, poor documentation of direct risks to human health and the ecosystem, a dearth of new U.S.-developed dredging technology, and legal limitations on the direct use of foreign technology. The report concluded that periodic reviews of site assessment procedures and cleanup technologies were needed and that management decisions should be based on a comparison of risks, costs, and benefits to both the environment and public health.

The 1989 NRC report enhanced public understanding of the widespread and important, but poorly quantified, problem of contaminated sediments and suggested how it might be addressed. The report assisted several states, the Environmental Protection Agency (EPA), the National Oceanic and Atmospheric

Administration (NOAA), and the U.S. Army Corps of Engineers (USACE) to address the issue of contaminated sediments in the context of other responsibilities. The U.S. Congress responded by mandating, in the Water Resources Development Act of 1992 (P.L. 102-580), an inventory of contaminated sediment sites.

There are four principal reasons to manage contaminated marine sediments: (1) to identify and clean up threats to public health and wildlife; (2) to meet water and environmental quality standards; (3) to identify and clean up sites that have the potential to cause wider environmental harm; and (4) to ameliorate dredging controversies, particularly concerning the designation of disposal sites for contaminated dredged material. A strategy for achieving these objectives must strike a balance among various risks and among risks, costs, and benefits. Choices that must be made from a wide range of tactics are hampered by substantial uncertainties. The present study is an attempt to assist in the decision making and to address the key management and technology issues associated with the remediation of contaminated marine sediments.

SCOPE OF THE STUDY

After discussions with the EPA, NOAA, USACE, and the U.S. Navy, the NRC convened a committee under the auspices of the Marine Board to assess the nation's capability for remediating contaminated marine sediments and to chart a course for the development of management strategies. The objectives of the study were:

- to assess the best management practices and current and emerging remediation technologies that have been tried for reducing adverse environmental impacts of contaminated sediments. These approaches include biological, chemical, and physical methods, such as removal technologies, in situ and ex situ treatment, containment (including capping), and natural recovery. Methods were to be reviewed with regard to scientific and engineering feasibility, practicality, cost, efficiency, and effectiveness.
- to appraise interim control measures for contaminated sediment sites. Interim control methods can be technology-based (e.g., systems that halt the deposition or spread of contaminants) or management-oriented (e.g., controlling other uses of contaminated areas). Interim control measures were to be identified and appraised to determine their applicability to classes of problems, their affordability, and their practicality.
- to examine how information about risks, costs, and benefits can be used to guide decision making concerning the management of contaminated sediments
- to assess existing knowledge and to identify research that is critical for enhancing the use of existing technologies in contaminated sediment management and in developing new technologies

PREFACE   ix

The committee determined that the effectiveness of contaminated sediment management practices and remediation technologies is influenced by a number of external factors, including laws and regulations, site assessment methods, and efforts to control the source(s) of contamination. Therefore, the committee judged it necessary to include these topics in its assessment, but only to the extent that they support best management practices. Evaluating the significance of natural spatial and temporal variations and identifying "clean" versus "contaminated" sediments are outside the purview of this study. Also, a detailed comparison of the biogeochemical and biological "availability" of contaminants with a concentration-driven process is beyond the scope of this report. These are, however, important topics that might be addressed elsewhere.

The study was carried out by a carefully constituted committee and staff. Committee members were selected to ensure a wide range of expertise and to include a broad spectrum of viewpoints. Members represented the fields of coastal, geotechnical, and systems engineering; site remediation and bioremediation; port engineering and operations; aquatic toxicology; physical, chemical, geological, and biological oceanography; geology; environmental law and policy; and economics. (Biographies of the committee members are provided in Appendix A.) In keeping with NRC policy, potential biases that might accompany expertise vital to the study were not excluded.

There is no universally accepted definition of a "contaminated" sediment. The 1989 NRC report defined the term to mean a sediment that contains chemical concentrations that pose a known or suspected threat to the environment or human health. The methods for determining if a risk exists are imperfect, so the "acceptable" level of contamination is subject to debate. Regulatory agencies, from the local to the international level, have adopted or are producing both qualitative and quantitative definitions of contaminated sediments. For purposes of this report, the committee assumed that methodological thresholds for determining when contamination exists are available and are used. The committee did not assess the adequacy of standard practices beyond pointing out how they may influence risk management.

STUDY METHOD

The full committee met seven times over a three-year period. The committee reviewed relevant reports and was briefed on federal activities related to contaminated sediments. Information was solicited from expert researchers and practitioners from federal, regional, state, and local government agencies; port authorities; industry; and public interest groups. The committee also visited the U.S. Army Engineer, Waterways Experiment Station (WES) in Vicksburg, Mississippi, where the committee was briefed on research activities, and to the Port of Tacoma in Washington, where the committee solicited expert testimony regarding an ongoing dredging and remediation project. The committee also held workshops on sediment removal and remediation technologies (Thoma, 1994) and on interim control measures.

In addition to the full committee meetings, various committee members developed particular aspects of the report either on their own or by working in small groups. Two committee members prepared a review of the regulatory framework for contaminated sediments (Appendix B), while others developed the case histories of six ongoing or recently completed remediation projects (summarized in Appendix C). Committee members with special expertise prepared primers on the application of two decision-making tools to improve contaminated sediments management (Appendix D and Appendix E).

REPORT ORGANIZATION

The audience for this report includes federal, state, and local government agencies; U.S. Congress and congressional staff; policymakers and project managers; members of the technical community associated with the various aspects of the remediation of contaminated sediments; and other members of the marine or coastal community, including the general public, who have a stake in the decision-making process.

Chapter 1 outlines the forces driving the remediation of contaminated sediments, the risk management process, and the unique challenges to be overcome—all factors that affect the choice of management techniques and technologies. Chapter 2 describes a conceptual management approach to the problem, from the identification of a contaminated site through the long-term monitoring of project results, as well as tools for assessing trade-offs among risks, costs, and benefits that can improve decision making.

In chapters 3, 4, and 5, specific topics are examined with the aim of enhancing the prospects for success. Chapter 3 discusses two important influences on decision making—regulatory realities and stakeholder interests—that must be mastered by project proponents. Chapter 4 describes how proper attention to site-specific considerations, including source control and site assessment, can support cost-effective management. The heart of the report is Chapter 5, which contains an assessment of interim and long-term controls and technologies on the basis of maturity, applicability, effectiveness, limitations, cost, and research needs.

Chapter 6 synthesizes the information and analyses in the previous chapters and presents conclusions and recommendations. The appendices include a review of the regulatory framework for contaminated sediments (Appendix B), a summary of the case studies (Appendix C), a primer on the use of cost-benefit analysis to improve management (Appendix D), and a description of decision analysis and its application in a test case (Appendix E).

ACKNOWLEDGMENTS

The committee wishes to thank the many individuals who contributed their time and effort to this project, whether in the form of presentations at meetings, correspondence, or telephone calls. Invaluable assistance was provided to both the committee and the Marine Board staff by representatives of federal and state agencies as well as private companies in various sectors.

In particular, the committee acknowledges the support of the following individuals: Sabine Apitz, U.S. Navy, Naval Command, Control and Ocean Surveillance Center; Charles C. Calhoun, U.S. Army Engineer Waterways Experiment Station (WES); Miles Croom, NOAA, National Marine Fisheries Service; Robert Engler, U.S. Army Engineer WES; Kenneth Hood, EPA, Office of Research and Development; Evie Kalketenidou, Maritime Administration (MARAD), Office of Port and Intermodal Development; Daniel Leubecker, MARAD, Office of Technology Assessment; Frank Manheim, U.S. Geological Survey (USGS), Branch of Atlantic Marine Geology; Janet Morton, USGS, Office of Energy and Marine Geology; Anna Palmisano, U.S. Navy, Office of Naval Research, Biological Science and Technology Program; Carl Sobremisana, MARAD, Office of Port and Intermodal Development; Betsy Southerland, EPA, Office of Science and Technology (OST); Mark Sprenger, EPA, Office of Emergency and Remedial Response; Craig Vogt, EPA, Office of Wetlands, Oceans, and Watersheds; Larry Zaragoza, EPA, Office of Solid Waste and Emergency Remedial Response; and Christopher Zarba, EPA, OST.

For assistance with technical aspects of the report, special thanks also go to Steve Garbaciak and James Hahnenberg of the EPA; Dan Averett, James Clausner, Norm Francingues, C.R. Lee, Jan Miller, Michael Palermo, and Joe Wilson of the USACE; Greg Hartman, Greg Hartman Associates; Ancil Taylor, Bean Dredging Corporation; and Ian Orchard and Carol Ancheta, Environment Canada.

Finally, the chairmen recognize members of the committee, not only for their hard work during meetings and in reviewing drafts of this report but also for gathering information and writing sections of the report.

REFERENCES

Maritime Administration (MARAD). 1994. A Report to Congress on the Status of the Public Ports of the United States, 1992–1993. MARAD Office of Ports and Domestic Shipping. Washington, D.C.: U.S. Department of Transportation.

National Research Council (NRC). 1989. Contaminated Marine Sediments: Assessment and Remediation. Washington, D.C.: National Academy Press.

Thoma, G. 1994. Summary of the Workshop on Contaminated Sediment Handling, Treatment Technologies, and Associated Costs held April 21–22, 1994. Background paper prepared for the Committee on Contaminated Marine Sediments, Marine Board, National Research Council, Washington, D.C.

Contents

EXECUTIVE SUMMARY		1
1	THE CHALLENGE	15
	Driving Forces for Remediation, 16	
	Risk Management Process, 22	
	Unique Challenges Posed by Contaminated Sediments, 23	
	Summary, 28	
	References, 29	
2	MAKING BETTER DECISIONS: A CONCEPTUAL MANAGEMENT APPROACH	30
	Conceptual View of the Management of Contaminated Sediments, 30	
	Trade-Offs in Risks, Costs, and Benefits, 34	
	Summary, 42	
	References, 43	
3	FORCES INFLUENCING DECISION MAKING	44
	Regulatory Realities, 46	
	Stakeholder Interests, 52	
	Summary, 60	
	References, 61	
4	SITE-SPECIFIC CONSIDERATIONS	62
	Sources of Contamination, 62	
	Contaminant Transport and Availability, 64	

Site Assessment: Approach, Methods, and Procedures, 67
Summary, 77
References, 77

5   INTERIM AND LONG-TERM TECHNOLOGIES AND CONTROLS   80

Introduction, 80
Interim Controls, 84
Technologies for In Situ Management, 91
Sediment Removal and Transport Technologies, 104
Technologies for Ex Situ Management, 116
Evaluating the Performance of Technologies and Controls, 137
Research, Development, Testing, and Demonstration, 141
Comparative Analysis of Technology Categories, 142
References, 147

6   CONCLUSIONS AND RECOMMENDATIONS   154

Improving Decision Making, 155
Improving Remediation Technologies, 161
Improving Project Implementation, 168

APPENDICES

A   BIOGRAPHICAL SKETCHES OF COMMITTEE MEMBERS   175
B   REGULATORY FRAMEWORK FOR THE MANAGEMENT AND REMEDIATION OF CONTAMINATED MARINE SEDIMENTS   181
C   CASE HISTORIES OF REPRESENTATIVE REMEDIATION PROJECTS   225
D   USING COST-BENEFIT ANALYSIS IN THE MANAGEMENT OF CONTAMINATED SEDIMENTS   239
E   USING DECISION ANALYSIS IN THE MANAGEMENT OF CONTAMINATED SEDIMENTS   257

Boxes, Figures, and Tables

BOXES

2-1 Evaluating Sediment Contamination: Effects-Based Testing and Sediment Quality Criteria, 38
4-1 Basic Tenets of Site Assessment, 69
5-1 Importance of Cost in Technology Assessment, 81
5-2 Process of Defining a Remediation System, 86
5-3 Selecting Ex Situ Controls, 117
D-1 Simplified Examples of Cost-Benefit Calculations, 250

FIGURES

1-1 Regulation of contaminated sediments, 17
1-2 Volume and costs of dredging by the USACE and industry, 1963 to 1994, 20
2-1 Conceptual overview of the management of contaminated sediments, 31
2-2 Conceptual illustration of the trade-offs involved in cost-benefit analysis, 40
4-1 Conceptual site assessment protocol, 68
5-1 Process of defining a remediation system, 83
5-2 Remediation technologies subsystem structure, 85
5-3 Conceptual illustration of containment, disposal, and natural recovery technologies, 131
D-1 Conceptual illustration of the trade-offs involved in cost-benefit analysis, 242

D-2 Example of cost-benefit analysis with discrete projects, 245
D-3 Costs and benefits of reducing body burden, 246
E-1 Predicted average PCB concentration as a function of area dredged (to a depth of 1 meter), assuming sediments are dredged in order of decreasing PCB concentration, 269
E-2 Influence diagram of a test case, 273
E-3 Expected values of alternative dredged-volume decisions, 278
E-4 Dredged-volume decision analysis, 279
E-5 Expected values of alternative dredged-value decisions with modified model parameters, 281
E-6 Effect of modified model parameters on maximum dredged volume decision, 282
E-7 Switchover analysis for the *annual resource damage cost,* 283

TABLES

S-1 Comparative Analysis of Technology Categories, 13
1-1 Time Lapse between Identification of a Problem and Implementation of a Solution: Examples from Six Case Histories, 26
5-1 Natural Recovery, 93
5-2 In-Place Capping, 96
5-3 Immobilization (solidification/stabilization), 98
5-4 In Situ Chemical Treatment, 98
5-5 In Situ Bioremediation, 101
5-6 Soil Washing and Physical Separation, 119
5-7 Chemical Separation and Thermal Desorption, 122
5-8 Immobilization, 124
5-9 Thermal and Chemical Destruction, 126
5-10 Ex Situ Bioremediation, 128
5-11 Confined Disposal Facility, 133
5-12 Contained Aquatic Disposal, 135
5-13 Landfills, 138
5-14 Qualitative Comparison of the State of the Art in Remediation Technologies, 143
5-15 Comparative Analysis of Technology Categories, 144
B-1 Interrelationships of Sediment Regulatory Authorities in Selected Scenarios, 215
C-1 Selection and Evaluation Criteria for Six Case Histories, 226

Acronyms

ARAR	applicable or relevant and appropriate regulatory requirement
ARCS	Assessment and Remediation of Contaminated Sediments
CAD	contained aquatic disposal
CDF	confined disposal facility
CERCLA	Comprehensive Environmental Response, Cleanup, and Liability Act
CFR	Code of Federal Regulations
CRADA	cooperative research and development agreement
CWA	Clean Water Act
CZMA	Coastal Zone Management Act
D&D	dredging and disposal (or placement)
DMMP	dredged material management plan
DOE	Department of Energy
DOI	Department of the Interior
EA	environmental assessment
EIS	environmental impact statement
EPA	Environmental Protection Agency
ESA	Endangered Species Act
FDA	Food and Drug Administration
FONSI	finding of no significant impact

GDP	gross domestic product
GPS	global positioning system
LA	load allocation
MBDS	Massachusetts Bay Disposal Site
MCY	millions of cubic yards
MPRSA	Marine Protection, Research and Sanctuaries Act
MT	metric tons
NAAQS	national ambient air quality standards
NEPA	National Environmental Policy Act
NOAA	National Oceanic and Atmospheric Administration
NPL	National Priorities List
NRC	National Research Council
NSI	National Sediment Inventory
OMC	Outboard Motor Corporation
OST	Office of Science and Technology (EPA)
PCB	polychlorinated biphenyl
RCRA	Resource Conservation and Recovery Act
RHA	Rivers and Harbors Act
ROD	record of decision
R&D	research and development
SARA	Superfund Amendments and Reauthorization Act
SITE	Superfund Innovative Technology Evaluations
SQC	sediment quality criteria
TCLP	toxic characteristics leaching procedure
TMDL	total maximum daily load
USACE	U.S. Army Corps of Engineers
USGS	U.S. Geological Survey
WES	U.S. Army Engineer, Waterways Experiment Station
WLA	waste load allocation
WRDA	Water Resources Development Act

Contaminated Sediments in Ports and Waterways

Port of Mobile (Overleaf)

Dredging enables ports to maintain adequate depths in harbors and channels and thereby attract commercial shipping, which provides goods, jobs, and other benefits for area residents. The photograph shows a hydraulic dredge drawing up sediment from a shipping channel in Mobile, Alabama.

Photograph courtesy of U.S. Army Corps of Engineers

Executive Summary

There is no simple solution to the problems created by contaminated marine sediments,[1] which are widespread in U.S. coastal waters and can pose risks to human health, the environment, and the nation's economy. Marine sediments are contaminated by chemicals that tend to sorb to fine-grained particles; contaminants of concern include trace metals and hydrophobic organics, such as dioxins, polychlorinated biphenyls (PCBs), and polyaromatic hydrocarbons. Contamination is sometimes concentrated in "hot spots" but is often diffuse, with low to moderate levels of chemicals extending no more than a meter into the seabed but covering wide areas. Approximately 14 to 28 million cubic yards of contaminated sediments must be managed annually, an estimated 5 to 10 percent of all sediments dredged in the United States.

The many challenges to be overcome in managing contaminated sediments include an inadequate understanding of the natural processes governing sediment dispersion and the bioavailability of contaminants; a complex and sometimes inconsistent legal and regulatory framework; a highly charged political atmosphere surrounding environmental issues; and high costs and technical difficulties involved in sediment characterization, removal, containment, and treatment. The need to meet these challenges is urgent. The presence of contaminated sediments poses a barrier to essential waterway maintenance and construction in many ports, which support approximately 95 percent of U.S. foreign trade. The management

---

[1] For purposes of this report, contaminated marine sediment is defined as containing chemical concentrations that pose a known or suspected threat to the environment or human health.

of these sediments is also an issue in the remediation[2] of an estimated 100 marine sites targeted for cleanup under the Comprehensive Environmental Response, Cleanup, and Liability Act (CERCLA) (P.L. 96-510), commonly known as Superfund, as well as in the cleanup of many other near-shore contaminated sites.

The Committee on Contaminated Marine Sediments was established by the National Research Council under the auspices of the Marine Board to assess the nation's capability for remediating contaminated marine sediments and to chart a course for the development of management strategies. In the committee's view, cost-effective management of contaminated marine sediments will require a multifaceted campaign as well as a willingness to innovate. The committee determined that a systematic, risk-based approach incorporating improvements to current practice is essential for the cost-effective management of contaminated marine sediments. The committee identified opportunities for improvement in the areas of decision making, project implementation, and interim and long-term controls and technologies, as outlined in this summary. Although the study focused on evaluating management practices and technologies, the committee also found it essential to address a number of tangentially related topics (e.g., regulations, source control, site assessment) because problems in these areas can impede application of the best management practices and technologies.

As part of the three-year study, the committee compiled six case histories of recent or ongoing contaminated sediments projects, visited one of those sites, analyzed the relevant regulatory framework in depth, held separate workshops on interim controls and long-term technologies, and examined in detail how various decision-making approaches can be applied in the contaminated sediments context. The committee also examined the application of decision analysis in contaminated sediments management.

IMPROVING DECISION MAKING

Decision-Making Tools

Contaminated sediments can best be managed if the problem is viewed as a system composed of interrelated issues and tasks. Systems engineering and analysis are widely used in other fields but have not been applied rigorously to the management of contaminated sediments. The overall goal is to manage the

---

[2] For purposes of this report, sediment management is a broad term encompassing remediation technologies as well as nontechnical strategies. Remediation refers generally to technologies and controls designed to limit or reduce sediment contamination or its effects. Controls are practices, such as health advisories, that limit the exposure of contaminants to specific receptors. Technologies include containment, removal, and treatment approaches. Treatment refers to advanced technologies that remove a large percentage of the contamination from sediment.

system in such a way that the results are optimized. In particular, a systems approach is advisable with respect to the selection and optimization of interim and long-term controls and technologies. Although unlimited time and money would make remediation of any site feasible, resource limitations demand that trade-offs be made and that solutions be optimized.

A fundamental aspect of the committee's recommended approach is the delineation of the trade-offs among risks, costs, and benefits that must be made in choosing the best course of action among multiple management alternatives. A number of decision-making tools can be used in making these trade-offs. Available tools include risk analysis, cost-benefit analysis, and decision analysis.

Cost-effective contaminated sediments management requires the application of risk analysis—the combination of risk assessment, risk management, and risk communication. Contaminated sediments are considered a problem only if they pose a risk that exceeds a toxicological benchmark. In its most elemental form, risk assessment is intended to determine whether the chemical concentrations likely to be encountered by organisms are higher or lower than the level identified as causing an unacceptable effect. The "acceptable risk" needs to be identified, quantified, and communicated to decision makers, and the risk needs to be managed. First, management strategies need to be identified that can reduce risk to an acceptable level. Second, remediation technologies need to be identified that can reduce the risk associated with contaminants to acceptable levels within the constraints of applicable laws and regulations. Third, promising technologies need to be evaluated within the context of the trade-offs among risks, costs, and benefits, a difficult task given the uncertainties in risk and cost estimates. The next step is risk communication, when the trade-offs are communicated to the public.

At present, risk analysis is not applied comprehensively in contaminated sediments management. Risks are usually assessed only at the beginning of the decision-making process to determine the severity of the in-place contamination; the risks associated with removing and relocating the sediments or the risks remaining after the implementation of solutions are not evaluated. The expanded application of risk analysis would not only inform decision makers in specific situations but would also provide data that could be used in the selection and evaluation of sediment management techniques and remediation technologies.

Cost-benefit analysis can also be useful for evaluating proposed sediment management strategies. Although risk assessments may provide information about the exposure, toxicity, and other aspects of the contamination, they may result in a less-than-optimum allocation of resources unless additional information is considered. For example, a given concentration of contaminants at a particular site might be toxic enough to induce mortality in a test species, but this information alone does not indicate the spending level that would be justified for cleanup. Cost-benefit analysis combines risk and cost information to determine the most efficient allocation of resources. The basic principle of cost-benefit analysis is that activities should be pursued as long as the overall benefit to society exceeds the social cost. The difficulty lies in the measurement of the benefits and costs, or, more to the point, the projection of what they will be, before a strategy is implemented.

Cost-benefit analysis is not applied widely in contaminated sediments management. It is generally carried out only for major new navigational dredging projects, and the analyses are usually narrow in scope. Cost-benefit analysis could be used in many cases to help identify the optimum solution in which the benefits outweigh the costs (i.e., to maximize benefits for a given cost or to minimize costs for a given level of benefits). The costs and benefits involved in contaminated sediments management are difficult to calculate and cannot be measured precisely, but cost-benefit analysis may be worth the effort; comprehensive cost-benefit analysis may be warranted in very expensive or extensive projects. Informal estimates or cost-effectiveness[3] analyses may suffice in smaller projects.

As the demand for the remediation of contaminated sediments grows, and as costs and controversies multiply, decision makers need to be able to use information about risks, costs, and benefits that may be controversial and difficult to evaluate, compare, or reconcile. One approach that could help meet this need is decision analysis, a computational technique that makes use of both factual and subjective information in the evaluation of the relative merits of alternative courses of action. Decision analysis involves gathering certain types of information about a problem and selecting a set of alternative solutions to be evaluated. The evaluation is used to determine and assess possible outcomes for each alternative. The outcomes are rated, and the results are used to develop a strategy that offers the best odds for successful risk management.

Formal decision analysis is not yet widely used in the management of contaminated sediments. The committee examined this technique using a test case and determined that applications of decision analysis may be particularly timely now, because recent advances in computer hardware and software make it possible to perform such analyses in ways that are user friendly and interactive. Decision analysis could be especially valuable because it can accommodate more variables (including uncertainty) than techniques such as cost-benefit analysis that measure single outcomes. Decision analysis can also serve as a consensus-building tool by enabling stakeholders to explore various elements of the problem and, perhaps, find common ground. However, because decision analysis is technical in design and involves complex computations, it is probably worth the effort only in highly contentious situations in which stakeholders are willing to devote enough time to become confident of the usefulness of the approach.

---

[3] Cost effectiveness is defined here as a measure of tangible benefits for money spent.

Regulatory Framework

Few aspects of sediment handling, treatment, or containment are unregulated at the federal, state, or local level, but the regulatory approach is inconsistent, primarily because the applicable laws were originally written to address issues other than contaminated marine sediments. As a result, the current laws and regulations affecting contaminated sediments can impede efforts to implement the best management practices and achieve efficient, risk-based, and cost-effective solutions. This is a shortcoming of the governing statutes, not a criticism of regulatory agencies charged with implementing them. The timeliness of decision making is also an issue, given that it typically takes years to implement solutions to contaminated sediments problems. In the committee's case histories, the delay between the discovery of a problem and the implementation of a solution ranged from approximately 3 to 15 years.

At least six comprehensive acts of Congress, with implementation responsibilities spread over seven federal agencies, govern sediment remediation or dredging operations in settings that range from the open ocean to the freshwater reaches of estuaries and wetlands. When environmental cleanup is the driving force, the relevant federal laws include Superfund; the Resource Conservation and Recovery Act (RCRA) (P.L. 94-580); and Section 115 of the Clean Water Act (CWA) (formerly the Federal Water Pollution Control Act [P.L. 80-845]). When navigational dredging is the issue, the applicable statutes are likely to be the CWA; the Rivers and Harbors Act of 1899 (P.L. 55-525); the Marine Protection, Research and Sanctuaries Act (MPRSA, commonly known as the Ocean Dumping Act) (P.L. 92-532); and the Coastal Zone Management Act (P.L. 92-583). In addition, states also exercise important authority related to water quality certification and coastal zone management. In some cases, local laws may also apply. To complicate matters further, federal, state, and local authorities often overlap.

The principal federal agencies involved are the Environmental Protection Agency (EPA), which is responsible for implementing Superfund and has major site designation, regulation development, and veto responsibilities under the CWA and MPRSA; the National Oceanic and Atmospheric Administration, which assesses the potential threat of Superfund sites to coastal marine resources and exercises significant responsibilities for research, under the MPRSA, and review and comment, under CWA and MPRSA; and the U.S. Army Corps of Engineers (USACE), which assists in the design and implementation of remedial actions, under Superfund, and has responsibilities for dredged material, under the CWA, MPRSA, and Rivers and Harbors Act. The federal navigational dredging program is the joint responsibility of the EPA and USACE; the EPA regulates disposal, whereas USACE handles the dredging.

The committee identified several areas of the current regulatory framework in which changes might be beneficial. For example, the CWA, the MPRSA, and

Superfund use different approaches for evaluating remedial alternatives, but none fully considers either the risks posed by contaminated marine sediments or the costs and benefits of various solutions. The MPRSA requires biological testing of dredged material to determine its inherent toxicity but does not fully consider site-specific factors that may influence the exposure of organisms in the receiving environment, meaning that, at best, risk is considered only indirectly and the actual impact is approximated. Although the CWA procedures, which consider chemical and physical as well as biological characteristics in assessing whether the discharge of dredged material will cause unacceptable adverse impacts, are not risk-based, at least they do not specify rigid pass-fail criteria. They are geared to identification of the least environmentally damaging, implementable alternative. The Superfund remedial action program addresses risks and costs to some degree—an exposure assessment (but not a full risk analysis) is required to assess in-place risks; remedial alternatives are identified based on their capability of reducing exposure risks to an acceptable level; and the final selection involves choosing the most cost-effective solution. However, there are no risk-based cleanup standards for underwater sediments. Insufficient attention to risks, costs, and benefits impedes efforts to reach technically sound decisions and manage sediments cost-effectively.

Similar inattention to risk is evident in the permitting processes for sediment disposal. It is currently necessary to secure different types of permits for the placement of sediments in navigation channels or ocean waters as part of the construction of land or containment facilities (under the Rivers and Harbors Act), the dumping of sediments in the ocean (under the MPRSA), the discharge of sediments in inland waters or wetlands (CWA), and the containment of contaminated sediments on land (RCRA). In addition, different regulations come into play depending on whether sediments are removed during navigational dredging (CWA or MPRSA) or are excavated for environmental remediation (Superfund). The committee can see little technical justification for the differential regulation of contaminated sediments, given that neither the location of the aquatic disposal site (freshwater versus saltwater) nor the reason for dredging (navigational dredging versus environmental remediation) necessarily affects the risk posed by the contamination. The regulatory regime does not adequately address risk; instead it focuses rigidly on the nature of the activities to be carried out. This problem has been eased in some instances by the interpretation of regulations based on the intent of the underlying statute(s).

Systematic, integrated decision making can also be undermined by dredging regulations governing cost allocation and cost-benefit analysis. The federal government pays for most new-work dredging and all maintenance dredging but not for sediment disposal, except in open water. The local sponsors of federal navigation projects bear the burden of identifying, constructing, operating, and maintaining dredged material disposal sites, under the "project cooperation agreement" of the Water Resources Development Act (WRDA) of 1986 (P.L. 99-662). Because project sponsors must pay for disposal on land, whereas open-water disposal is paid for by the federal government as a component of dredging costs, the WRDA provision creates a strong preference for open-water disposal. Furthermore, a local sponsor bearing the full burden of disposal costs has little incentive to seek out opportunities for the beneficial uses of dredged material (discussed in the next section). The cost of making use of dredge material adds to the project cost and may benefit only third parties. This inconsistent approach to cost sharing can lead to the economically irrational allocation of scarce societal resources. Additional inconsistencies are introduced in the area of cost-benefit analysis. As noted earlier, costs and benefits must be weighed for new dredging projects but not for the maintenance dredging of existing channels or for the disposal of dredged material.

IMPROVING PROJECT IMPLEMENTATION

Stakeholder Interests

Contaminated sediments are not managed in a political or social vacuum. Most contaminated sediments sites are located in highly populated areas near the Great Lakes or the oceans. The nature of these sites virtually ensures that complicated ecological situations and difficult technical problems will have to be accommodated along with complex political circumstances involving multiple resource users and interest groups. Stakeholders include port managers and transportation officials who have strong economic reasons for dredging; federal, state, and local regulators responsible for protecting natural resources and enforcing regulations; and environmental groups, local residents, fishermen, and other marine resource users who are concerned about public health and natural resources. The successful management of contaminated sediments must respond to all dimensions of the problem: ecological, technical, social, and political.

The committee determined that remediation and disposal projects need strong proponents and that the identification and timely implementation of effective solutions depend heavily on how project proponents interact with stakeholders, who often have different perspectives on the problem and proposed solutions. Because any participant in the decision-making process can block or delay remedial action, project proponents need to identify all stakeholders and build a consensus among them. The development of a consensus can be fostered by the use of various tools, including mediation, negotiated rule making, collaborative problem solving, and effective communication of risks.

Stakeholder acceptance of contaminated sediments management projects can be fostered by the reuse of dredged material. Dredged material has been used for many purposes, including the creation of thousands of islands for seabird nesting, landfills for urban development, and wetlands, as well as for beach nourishment and shoreline stabilization. The policy focus and most of the experience to date have concerned the use of clean materials, but some contaminated sediments can also be used safely for certain beneficial purposes. Reuse can provide alternatives to increasingly scarce disposal sites while also making management plans more attractive, or at least palatable, to stakeholders. Some contaminated sediment sites have been successfully transformed into wetlands, and productive USACE research is under way on the safe use of contaminated sediments for "manufacturing" topsoil and landfill covers. However, funding for this type of research is limited, and technical guidelines have yet to be developed. Other barriers include the USACE policy of selecting lowest-cost disposal options with little regard to the possibilities of beneficial use and the uncertainties about whether the incremental costs of beneficial use should be borne by the project proponent or the beneficiary.

Source Control

Because accumulations of sediments interfere with deep-draft navigation, ports have no alternative but to dredge periodically in order to remain economically viable. If the sediments to be dredged are contaminated, then ports become responsible for both sediment disposal and any necessary remediation, even though they have no control over the source of the contamination. Upstream generators of contaminants often cannot be identified or held accountable, leaving ports to manage a problem that is not of their making. This responsibility could be shared by states (when states do not already operate or oversee port agencies), which benefit economically from dredging and already engage in watershed management. Under the CWA (Section 303), the EPA and the states set total maximum daily loads for waterway segments and develop load allocations for pollution sources in an effort to control water pollution. This approach could be readily expanded to address sources of sediment contamination. In addition, government regulators and ports could use all available legal and enforcement tools for ensuring that polluters bear a fair share of cleanup costs.

Site Characterization

Accurate site characterization is essential to the cost-effective management of contaminated sediments. Site assessments need to be sufficiently comprehensive and accurate to ensure that the contamination is well defined both chemically and geographically. Inaccuracies and incompleteness can leave areas of unidentified contamination that pose continuing unmanaged risks. Another compelling argument for accurate site assessment is the need to control remediation costs; precise site definition is necessary to facilitate removal of only those sediments

EXECUTIVE SUMMARY 9

that are contaminated, thus controlling the volume of material that requires expensive remediation. But the high cost of commonly used site characterization technologies (i.e., physical profiling and chemical testing) has limited the precise definition of either horizontal or vertical contaminant distributions, which may have led to the removal and "remediation" of large quantities of uncontaminated sediments at unnecessarily high costs.

Thus, the development and wide use of new or improved site characterization technologies that are less expensive than current methods would enhance the cost-effective management of contaminated sediment sites. One technology that may prove useful in the future is acoustic profiling,[4] which helps define the thickness and distribution of disparate sediment types. Because contaminants tend to be associated with fine-grained material, acoustic profiling may provide for cost-effective remote surveying of contaminated sediments, thereby increasing the precision and accuracy of site assessment. Additional research and development is needed, however. Sediment characterization may also be enhanced through the adaptation of chemical sensors now used in the assessment of soil and groundwater sites.

INTERIM AND LONG-TERM CONTROLS AND TECHNOLOGIES

The following is a brief assessment of the controls and technologies that are applicable to contaminated sediments. The section concludes with a comparative analysis reflecting the committee's overall judgments of the feasibility, effectiveness, practicality, and cost of each control and technology.

Interim Controls

Interim controls may prove helpful when sediment contamination poses an imminent hazard. Identification of an imminent hazard is usually a matter of judgment, but in general an imminent hazard exists when contamination levels exceed by a significant amount the sum of a defined threshold level plus the associated uncertainty. Administrative interim controls (e.g., signs, health advisories) have been used a number of times. Only two applications of structural interim approaches (e.g., thin caps) were identified by the committee, but additional structural approaches, such as the use of confined disposal facilities (CDFs) for temporary storage, appear promising. Few data are available concerning the effectiveness of interim controls because to date they have not been used often or evaluated in detail.

---

[4] Acoustic profiling involves high-resolution mapping of the acoustic reflectivity of sediments.

Long-Term Controls and Technologies

Technologies for remediating contaminated sediments are at various stages of development. Sediment-handling technologies are the most advanced, although benefits can be realized from improvements in the precision of dredging (and, concurrently, site characterization). The state of practice for in situ controls ranges from immature (e.g., bioremediation) to evolving (e.g., capping). Ex situ containment is commonplace. A number of existing ex situ treatment technologies can probably be applied successfully to treating contaminated sediments, but full-scale demonstrations are needed to determine their effectiveness. But these technologies are expensive, and it is not clear whether unit costs would drop significantly in full-scale implementation.

The cost of cleanup depends on the number of steps involved—the more handling required, the higher the cost—and the type of approach used. The costs of removing and transporting contaminated sediments (generally less than $15 to $20/yd^3) tend to be higher than costs of conventional navigational dredging (seldom more than $5/yd^3) but much lower than the costs of treatment (usually more than $100/yd^3). Volume reduction (i.e., removing only sediments that require treatment and entraining as little water as possible) will mean greater cost savings than increased production rates; improved site characterization coupled with precision dredging techniques hold particular promise for reducing volume. Treatment costs may also be reduced through pretreatment.

In situ management offers the potential advantage of avoiding the costs and potential material losses associated with the excavation and relocation of sediments. Among the inherent disadvantages of in situ management is that they are seldom feasible in navigation channels that are subject to routine maintenance dredging. In addition, monitoring needs to be an integral part of any in situ approach to ensure effectiveness over the long term.

Natural recovery is a viable alternative under some circumstances and offers the advantages of low cost and, in certain situations, the lowest risk of human and ecosystem exposure to sediment contamination. Natural recovery is most likely to be effective where surficial concentrations of contaminants are low, where surface contamination is covered over rapidly by cleaner sediments, or where natural processes destroy or modify the contaminants, so that contaminant releases to the environment decrease over time. A disadvantage of natural recovery is that the sediment bed is subject to resuspension by storms or anthropogenic processes. For natural recovery to be pursued with confidence, the physical, chemical, and hydrological processes at a site need to be understood adequately; however, no capability currently exists for completely quantifying chemical movements. Extensive site-specific studies may be required.

In situ capping promotes chemical isolation and may protect the underlying contaminated sediments from resuspension until naturally occurring biological degradation of contaminants has occurred. The original bed must be able to support the cap, suitable capping materials must be available to create the cap, and suitable hydraulic conditions (including water depth) must exist to permit placement of the cap and to avoid compromising the integrity of the cap. Changes in the local substrate, the benthic community structure, or the bathymetry at a depositional site may subject the cap to erosion. Improved long-term monitoring methods are needed. A regulatory barrier to the use of capping is the language of Superfund legislation (Section 121[b]), which gives preference to "permanent" controls. Capping is not considered by regulators to be a permanent control, but available evidence suggests that properly managed caps can be effective.

Neither in situ immobilization nor chemical treatment of contaminated sediments has been demonstrated successfully in the marine environment, although both concepts are attractive because they do not require sediment removal. Their application would be complicated by the need to isolate sediments from the water column during treatment, by inaccuracies in reagent placement, and by the need for long-term follow-up monitoring. Other constituents (e.g., natural organic matter, oil and grease, metal sulfide precipitates) could interfere with chemical oxidation. Immobilization techniques may not be applicable to fine-grained sediments with a high water content.

Biodegradation has been observed in soils, in groundwater, and along shorelines contaminated by a variety of organic compounds (e.g., petroleum products, PCBs, polyaromatic hydrocarbons, pesticides). However, the use of biodegradation in subaqueous and especially marine environments presents unresolved microbial, geochemical, and hydrological issues and has yet to be demonstrated.

When sediments must be moved for ex situ remediation or confinement, efficient hydraulic and mechanical methods are available for removal and transportation. Most dredging technologies can be used successfully to remove contaminated sediments; however, they have been designed for large-volume navigational dredging rather than for the precise removal of hot spots. Promising technologies offering precision control include electronically positioned dredge heads and bottom-crawling hydraulic dredges. The latter may also have the capability to dredge in depths beyond the standard maximum operating capacity. The cost effectiveness of dredging innovations can best be judged by side-by-side comparisons to technologies in current use.

Containment technologies, particularly CDFs, have been used successfully in numerous projects. A CDF can be effective for long-term containment if it is well designed to contain sediment particles and contaminants and if a suitable site can be found. A CDF can also be a valuable treatment or interim storage facility, allowing the separation of sediments for varying levels of treatment and, in some cases, beneficial reuse. Costs are reasonable; in some parts of the country it may be cheaper to reuse CDFs than to build new ones. Disadvantages of this technology include the imperfect methods for controlling contaminant release pathways. There is also a need for improved long-term monitoring methods.

Contained aquatic disposal (CAD) is applicable particularly to contaminated sites in shallow waters where in situ capping is not possible and to the disposal and containment of slightly contaminated material from navigation dredging. Although the methodology has been developed, CAD has not been widely used. Among the advantages of CAD are that it can be performed with conventional dredging equipment and that the chemical environment surrounding the cap remains unchanged. Disadvantages include the possible loss of contaminated sediments during placement operations. Improved tools are needed for the design of sediment caps and armor layers and for the evaluation of their long-term stability and effectiveness.

Scores of ex situ treatment technologies have been bench tested and pilot tested, and some warrant larger-scale testing in marine systems, depending on their applicability to particular problems. Chemical separation, thermal desorption, and immobilization technologies have been used successfully but are expensive, complicated, and only effective for treating certain types of sediments. Similarly, because of extraordinarily high unit costs, thermal and chemical destruction techniques do not appear to be near-term, cost-effective approaches for the remediation of large volumes of contaminated dredged sediment.

Ex situ bioremediation, which is not as far along in development as are other ex situ treatment approaches, presents so many technical problems that its application to contaminated sediments would be expensive. If these technical problems can be resolved, however, ex situ bioremediation has the potential, over the long term, for the cost-effective remediation of large volumes of sediments. Ex situ bioremediation is much more promising than in situ bioremediation because conditions can be controlled more effectively in a contained facility. The approach has been demonstrated on a pilot scale with some success, but complex questions remain concerning how to engineer the system.

Comparative Analysis of Controls and Technologies

Table S-1 summarizes the committee's overall assessment of the feasibility, effectiveness, practicality, and costs of controls and technologies. For each control and technology, the four characteristics were rated separately on a scale of 0 to 4, with 4 representing the best available (not necessarily the best theoretically possible) features. The effectiveness rating is an estimate of contaminant reduction or isolation and removal efficiency; scores represent a range of less than 90 percent to nearly 100 percent. The feasibility rating represents the extent of technology development, with 0 for a concept that has not been verified experimentally and 4 for a technology that has been commercialized. The practicality ranking reflects public acceptance; 0 means no tolerance for an activity and 4 represents widespread acceptance. The cost ranking is inversely related to the cost of using the control or

TABLE S-1 Comparative Analysis of Technology Categories

Approach	Feasibility	Effective	Practicality	Cost
INTERIM CONTROL				
Administrative	0	4	2	4
Technological	1	3	1	3
LONG-TERM CONTROL				
In Situ				
Natural recovery	0	4	1	4
Capping	2	3	3	3
Treatment	1	1	2	2
Sediment Removal and Transport	2	4	3	2
Ex Situ Treatment				
Physical	1	4	4	1
Chemical	1	2	4	1
Thermal	4	4	3	0
Biological	0	1	4	1
Ex Situ Containment	2	4	2	2
SCORING				
0	< 90%	Concept	Not acceptable, very uncertain	$1,000/yd^3
1	90%	Bench		$100/yd^3
2	99%	Pilot		$10/yd^3
3	99.9%	Field		$1/yd^3
4	99.99%	Commercial	Acceptable, certain	< $1/yd^3

technology (not including expenses associated with monitoring, environmental resource damage, or the loss of use of public facilities).

The overall pattern of the ratings underscores the need for trade-offs in the selection of technologies. No single approach emerges with the highest scores across the board, and each control or technology has at least one low or moderate ranking. In general, interim controls and in situ approaches are feasible and low in cost but less effective than the most practical ex situ approaches, which tend to be high in cost and complexity. Decisions about which approach is the most appropriate must be made on a project-by-project basis.

1

The Challenge

Sediment particles of mineral and organic matter accumulate in coastal waters as the result of physical, chemical, and biological processes, both natural and anthropogenic. Human activities can affect marine sediments by accelerating the rate of accumulation and introducing contamination. Many chemical contaminants have an affinity for fine-grained sediment particles. Contaminated sediments are widespread in U.S. coastal waters and have potentially far-reaching consequences to both public health and the environment (National Research Council [NRC], 1989a).

Industries located in or upstream of urban ports or industries that discharge wastes into waterways can be direct sources of contamination. Dense populations also contribute contaminants through sewage discharges, automobile emissions, and other waste-generating activities. Sediments can be contaminated by remote sources, such as stormwater runoff and suburban or agricultural effluents containing heavy metals, oil, pesticides, and fertilizers. Because estuaries have a natural tendency to trap sediment, contaminants from distant sources can be concentrated in already-stressed industrial harbors. Contaminants deposited from the atmosphere can be carried from sources even further afield. Recent studies have shown that about half of the metal contamination in the sediments of Long Island Sound may have come from atmospheric fallout (Cochran et al., 1993). Contamination sometimes concentrates in "hot spots" but is often diffuse, with low to moderate levels of chemicals less than a meter deep but covering wide areas.

Chemical contaminants associated with sediments can be considered toxic when they adversely affect living organisms. Submerged contaminated sediments may be in intimate contact with aquatic biota that may be affected adversely by, or serve as carriers of, contamination. In this way, contaminants pose a potential risk to coastal ecosystems and, primarily through consumption of fish and shellfish, to human health.

Management of contaminated sediments is a complicated problem.[1] At the technical level, controlling input is difficult because of the multiplicity of sources, and the wide dispersion of sediments by hydrodynamic and biological processes tends to expand the scope of cleanup operations. At the legal level, ports that may have no causal role in the contamination of sediments but must still dredge channels are faced with a number of hurdles, including identifying and paying for space for the placement of dredged material and many chemical, regulatory, political, and technological challenges.[2] Proper management of contaminated sediments is becoming more complicated because environmental concerns increasingly hinder the removal of sediments from economically critical shipping lanes and because growing numbers of contaminated sites are being identified for remediation.

DRIVING FORCES FOR REMEDIATION

Contaminated sediment becomes an issue when an environmental or human health risk is identified or when navigational needs require that contaminated sediment be dredged from shipping channels. Environmental risks may lead to the identification of human health risks and to limits on fishing or recreational uses of marine resources. The presence of contamination can make removing sediments that obstruct navigation in and around important ports very expensive. The choice of a remediation strategy is determined in large part by whether the driving force is environmental cleanup or navigational needs.

In addition to influencing the choice of remediation strategies, the driving force also affects which laws and regulations apply. At least seven federal agencies and six comprehensive Acts of the U.S. Congress influence remediation or dredging operations for managing contaminated sediments in settings that range from the open ocean to the inland and freshwater reaches of estuaries and wetlands (see Figure 1-1). If environmental cleanup is the driving force, applicable laws include the Comprehensive Environmental Response, Cleanup, and

---

[1] For purposes of this report, sediment management is a broad term encompassing remediation technologies as well as nontechnical strategies. Remediation refers generally to technologies and controls designed to limit or reduce sediment contamination or its effects. Controls are practices, such as health advisories, that limit the exposure of contaminants to specific receptors. Technologies include containment, removal, and treatment approaches. Treatment refers to advanced technologies that remove a large percentage of contamination from sediments.

[2] When referring to the final placement sites for dredged material, this report uses the terms of art established by applicable laws. Sediments are "dumped" in the open ocean (where the Marine Protection, Research and Sanctuaries Act applies) but "discharged" or "disposed of" in near-shore or inland waters (where the Clean Water Act applies). "Placement" is a generic term referring to all sites, both in the water and on land.

17

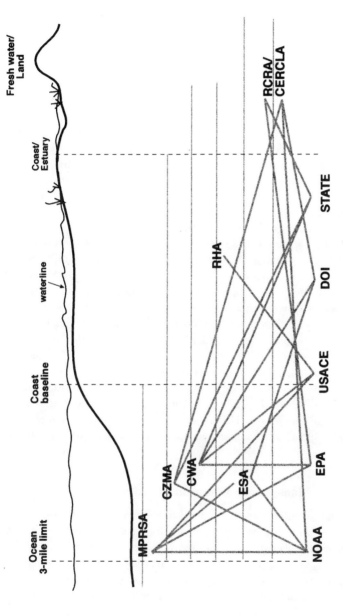

FIGURE 1-1 Regulation of contaminated sediments. The complexity of contaminated sediment regulation is depicted in this schematic diagram, which shows the locations to which various regulations apply, the interactions between regulations, and the responsible government agencies.

Note: USACE, U.S. Army Corps of Engineers; CWA, Clean Water Act; CZMA, Coastal Zone Management Act; DOI, Department of the Interior; EPA, Environmental Protection Agency; ESA, Endangered Species Act; MPRSA, Marine Protection, Research and Sanctuaries Act; NOAA, National Oceanic and Atmospheric Administration; RCRA/CERCLA, Resource Conservation and Recovery Act/Comprehensive Environmental Response, Cleanup, and Liability Act; RHA, Rivers and Harbors Act; state, any state government.

Liability Act (CERCLA), commonly known as Superfund (P.L. 96-510); the Resource Conservation and Recovery Act (RCRA) (P.L. 94-580); and Section 115 of the Clean Water Act (CWA) (originally called the Federal Water Pollution Control Act [P.L. 80-845 (1948)]). If navigation dredging is the issue, the applicable statutes are likely to be the CWA; the Rivers and Harbors Act of 1899 (P.L. 55-525); the Marine Protection, Research and Sanctuaries Act (MPRSA, also known as the Ocean Dumping Act) (P.L. 92-532); and the Coastal Zone Management Act (CZMA) (P.L. 92-583).

Three federal agencies are most active in contaminated sediment issues. The Environmental Protection Agency (EPA) is responsible for implementing Superfund and has major responsibilities and veto power for site designation and regulation development under the CWA and MPRSA. The National Oceanic and Atmospheric Administration is responsible for assessing the potential threat of Superfund sites to coastal resources, has significant research responsibilities under MPRSA, and has review obligations under both the CWA and MPRSA. The U.S. Army Corps of Engineers (USACE) assists in the design and implementation of remedial actions under Superfund and exercises primary responsibilities for permitting dredged material under the CWA, MPRSA, and Rivers and Harbors Act. The federal navigation dredging program is the responsibility of the EPA and USACE; the EPA addresses issues pertaining to disposal, and the USACE handles the dredging. Other federal, state, and local agencies have a hand in these matters as well. States are authorized to establish water quality standards within their jurisdictions and can block actions, such as sediment dredging or disposal, if they violate these standards. States also have the authority to review plans for consistency with coastal zone management plans. (Appendix B provides additional details on the regulatory framework.)

The overlapping jurisdictions of federal, state, and local authorities further complicate the situation, which is discussed further in the forthcoming section, Regulatory and Legal Challenges. The federal laws and regulations that apply to the handling and disposal of contaminated sediments are reviewed in detail in Appendix B.

Management of Natural Resources

Environmental cleanup, almost by definition, involves small volumes of highly contaminated sediment usually emanating from a known historical source and confined to well-defined areas. In environmental cleanup projects, the remediation strategy can be either in situ (i.e., in-place containment or treatment of the sediment) or ex situ (i.e., removal and disposal or treatment elsewhere). The removal of contaminated sediment for the sole purpose of cleanup as part of navigation projects has been permitted only in recent years in the United States. The USACE was given specific authorization under Section 312 of the Water Resources Development Act (WRDA) of 1990 (P.L. 101-640) to remove contaminated sediment outside the bounds of, but adjacent to, navigation channels. However, this apparently broad authority to clean up contaminated sediment in conjunction with federal navigation projects has not been used to date by the USACE for specific cleanup projects because of the inability to locate financially viable project sponsors and because of concerns about liability.

When sediment removal is not required for navigation, contaminated sediment may go unrecognized and the problem remain undefined until some event (e.g., routine water quality analysis) triggers recognition that the sediment may pose a risk to human health or the environment. Actual risk can be identified through a formal assessment (a process described in Chapter 2). Sometimes the response to this risk has an obvious and direct impact and economic consequences, such as restrictions on particular fisheries. In other cases, the response may be less visible but still significant in terms of impact, as when a site is designated in a Superfund "hazard ranking." Economic impact, as well as a high degree of risk, may make cleanup necessary.

The full extent of the need for environmental cleanup has not been quantified, but it is substantial. Approximately 100 marine sites[3] have been listed or proposed for inclusion on the National Priorities List (NPL) for long-term remedial action under Superfund, which addresses inactive or abandoned facilities that threaten public health or the environment. A national inventory[4] of contaminated sediment sites mandated by Congress in WRDA 1992 (P.L. 102-580) is under way. The EPA has designed the database and compiled the data and is expected to submit the first report to Congress in 1997.

Navigation Needs

Contaminated sediments usually accumulate slowly over large areas of the seafloor, but they can also accumulate very rapidly, especially in artificially deepened and confined areas, such as navigational channels and anchorages. Sediments in these areas must be dredged to maintain navigable waters. Navigation dredging typically involves the removal of large volumes of material over a large area that contains many different types of contaminants, albeit in low concentrations, from multiple, unidentifiable sources. In situ remediation strategies, such as leaving the contaminated sediment in place (i.e., allowing natural recovery to occur), may not be feasible in navigation channels. When navigation is the driving force, usually only ex situ techniques can be considered because the sediment must be relocated so the channel or harbor can be deepened or widened. In isolated instances, however, overdredging and capping the contaminated fraction of the dredged sediment within the navigation channel can be considered.

---

[3] This estimate includes Superfund sites adjacent to oceans and bays (L. Zaragosa, EPA, personal communication to Marine Board staff, October 1995).

[4] The inventory was released by the EPA for external review in July 1996.

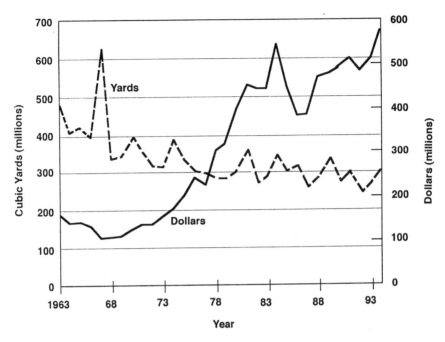

FIGURE 1-2 Volume and costs of dredging by the USACE and industry, 1963 to 1994. Estimates do not include disposal costs and are current, not constant, dollars.

Dredging is commonplace in the United States and is essential to many of the routine activities and services Americans have come to expect and demand (Maritime Administration, 1994). Figure 1-2 summarizes the volume (in millions of cubic yards [MCY]) and costs of dredging by the USACE and industry from 1963 to 1994 (USACE, 1995). Approximately 283 MCY of material, on the average, has been dredged annually in recent years from U.S. coastal and inland waters (1987 to 1994). Dredging and associated sediment disposal are expensive. The actual costs vary dramatically depending primarily on the nature (including contamination status) of the material to be dredged, the distance it must be transported for disposal, the number and nature of the required handling steps, the extent to which pre-disposal treatment is necessary, and the need for post-disposal monitoring. For major projects, environmental regulators require that all alternatives be explored before a decision to dredge is made, to ensure that a less costly or more environmentally acceptable alternative has not been overlooked.

Although the economic impact of *not* dredging sediment is difficult to quantify, there is no doubt that well maintained channels, ports, and harbors are essential if the United States is to continue to attract and retain commercial shipping (Interagency Working Group on the Dredging Process, 1994). Ports and harbors are essential to the nation's competitiveness in world markets. Approximately 95 percent of all U.S. foreign trade is waterborne and passes through U.S. ports (Maritime Administration, 1994). In 1992 the volume of waterborne foreign trade reached 897 million metric tons (MT) and was valued at $488 billion. It is expected that the value of imports and exports will increase from $488 billion in 1992 to $1.6 trillion in 2010, while increasing in volume from 897 million MT to 1.5 billion MT (U.S. Department of Transportation, 1994).

There are two types of navigation dredging: maintenance dredging and new-work (or new-construction) dredging. Maintenance dredging is carried out to maintain existing navigation services, whereas new-work dredging is intended to expand existing navigation channels or make them accessible to ships of deeper draft or to create new ones. Maintenance dredging is the more common of the two types. From 1987 to 1994, maintenance dredging in the United States moved, on average, approximately 238 MCY per year. This total includes dredging by the USACE on the inland waterway system and in federal channels of deep-draft ports, as well as dredging in other ports and by private parties. The amount of new-work dredging varies from year to year, depending on the commercial need for extended navigation facilities and the level of congressional appropriations. From 1987 through 1994, a total of 359 MCY was dredged for new construction. Most new-work dredging is associated with large federal projects. To qualify as a federal project—approved by U.S. Congress and/or under the management of the USACE—the benefits must be greater than the costs. Federal cost-sharing policies make a distinction between new-work dredging, for which the local sponsor must share the cost, and maintenance dredging, which is financed fully by the federal government through the Harbor Maintenance Trust Fund.[5]

New-construction dredging is motivated primarily by economics—that is, regional development pressures as well as the competitive position of a local port in relation to neighboring ports. Port upgrades also benefit the nation as a whole by supporting trade, an important element of the U.S. economy. Foreign trade now accounts for 20 percent of the gross domestic product (GDP), and this percentage is expected to grow in the future (Interagency Working Group on the Dredging Process, 1994). The combined economic impact of U.S. ports, port users, and public port capital expenditures is substantial. The demand for waterborne cargo initiates a chain of activity that contributes to the national economy. In 1992 U.S. ports handled approximately 2.9 billion MT of cargo, supported the employment of 15 million Americans, added $780 billion to the GDP and $523 billion to personal income, and contributed $210 billion in taxes to all levels of government (Maritime Administration, 1994).

---

[5] The trust fund is supplied by a tax levied on cargo passing through U.S. ports. The status of the fund was unclear as of late 1996; the U.S. Court of International Trade has ruled that the tax on exports is unconstitutional. The government, which claims the tax is actually a user fee, was expected to appeal the decision.

In summary, sustained U.S. economic growth is expected to depend increasingly on foreign trade and international commerce, most of which currently moves through the nation's ports. But ports cannot support economic growth without corresponding improvements in their capacity to accommodate an increase in commercial shipping as well as new classes of ships, some of which may have greater beam and deeper drafts than those found on today's vessels. Contaminated sediments and associated management difficulties can impede the expansion of navigational capacity and impose economic penalties.

RISK MANAGEMENT PROCESS

Contaminated marine sediments can pose risks to public health and the environment, and sound decisions about health and ecological risks must be based on formal assessments of those risks. The most elemental form of risk assessment is intended to determine whether the concentrations likely to be encountered by organisms are higher or lower than the level identified as causing an unacceptable effect. In this context, an effects assessment is a determination of the toxic concentration and the duration of exposure necessary to cause an effect of concern in a given species. Risk assessment as a decision-making tool is widely accepted in the scientific and engineering communities (NRC, 1983) and has been endorsed by the USACE for dredging operations (USACE, 1991). Risk is discussed in other reports of the NRC (1989b, 1993a,b, 1994a,b,c, 1995, and 1996), which address broad issues linking risk, science, and policy. Risk management is the evaluation, selection, and implementation of alternative methods of risk control.

Contaminated sediments are considered a problem only if they pose a risk above a toxicological benchmark, or acceptable level, which can be identified through a risk assessment.[6] Once the "acceptable risk" has been identified and quantified, a series of challenges in risk management become apparent. These challenges are outlined here to lay the groundwork for the analysis in forthcoming chapters.

First, management strategies must be identified that reduce risk to the benchmark value. The values currently used as benchmarks are imperfect in that they are based on inconsistent or incomplete applications of risk assessment principles (as discussed in chapters 2 and 3). Second, remediation technologies must be identified that can reduce the risk associated with contaminants to acceptable levels (sometimes known as "environmentally acceptable end-points"[7]) within

---

[6] The application of the risk assessment process to environmental or cleanup dredging has been summarized by the USACE (1991).

[7] An environmentally acceptable end-point is defined for soils as "a concentration of chemical(s) or test response(s) that is judged acceptable by a regulatory agency or other appropriate entity either by a standard or guideline, or which is derived using site-specific information" (Nakles and Linz, in

the constraints of applicable laws and regulations. Imminent health or environmental risks may call for prompt interim action and, later, more complete remediation. Where initial risk levels warrant some action but are not critically high, slower remediation tactics, such as natural recovery, may be appropriate. The capabilities of the various remediation technologies for reducing risk can only be estimated (see Chapter 5). Third, promising alternatives must be evaluated within the context of making trade-offs among risks, costs, and benefits. This is a difficult process, due in part to the uncertainties of risk and cost estimates. Fourth, the trade-offs must be communicated effectively to the stakeholders who have a say in the allocation of resources and an interest in ensuring that the decision-making process results in the successful resolution of the problem.

UNIQUE CHALLENGES POSED BY CONTAMINATED SEDIMENTS

Chemical Challenges

Marine sediments are contaminated by chemicals that tend to sorb to fine-grained particles, which offer a greater combined surface area for contaminant sorption than coarser particles (Gibbs, 1973; Moore et al., 1989). The contaminants of concern include trace metals and hydrophobic organics, such as dioxins, polychlorinated biphenyls (PCBs), and polyaromatic hydrocarbons. Metals bind to mineral surfaces or are present as sulfide precipitates. Because of the physiochemical state of the hydrophobic organics, they tend either to sorb to natural organic matter and fine clays or to be partitioned into a separate liquid phase, such as oil or coal tar. As a result, most highly contaminated sediments, regardless of the source of the contamination, tend to be fine-grained materials deposited in low-energy areas, which serve as sinks. The strong binding of contaminants with sediment, and their correspondingly slow release, suggest that risks to humans and the ecosystem, both lethal and sublethal, are linked to long-term rather than transitory exposure.

The accumulation of mixed contaminants complicates the selection of management strategies and treatment technologies for three reasons. First, opportunities for controlling the sources of contamination are limited, given that many different sources, some of them remote, may have contributed to the problem. Second, different types of contaminants must sometimes be treated in different ways. Third, a mixture of contaminants virtually guarantees that any treatment will leave behind untreated components. However, one particular contaminant is

---

press). As an example, it has been postulated that effective bioremediation can reduce hydrocarbon concentrations in soil to a level where they no longer pose an unacceptable risk to the environment or human health. It is believed that the remaining levels of hydrocarbons in the treated soil are no longer available to the environment or ecological and human receptors and represent an environmentally acceptable end-point.

usually the primary concern at a specific site, and the nature of this contaminant dictates the choice of remedial techniques.

Typical fine-grained contaminated sediments tend to have a relatively high water content and poor engineering qualities. Moreover, improper handling can remobilize the contaminants. Pore water containing dissolved contaminants may escape during dredging or transport.[8] In addition, small particles released into the water during handling have low settling rates and remain suspended in the water column where they are subject to wide dispersion. Special measures, such as silt curtains or water-tight bucket dredges, may be needed to limit the spread of resuspended contaminated sediments in some settings. Low settling rates can also complicate containment in a confined disposal facility (CDF); coagulating agents may be necessary to speed settling and reduce turbidity. But a percentage of fine-grained material and associated contaminants may remain suspended.

Regulatory and Legal Challenges

The regulations affecting contaminated sediments management are complicated. They were developed to implement a range of unrelated federal and state statutes dealing with issues, such as water quality and hazardous waste cleanup. As a result, the framework is inconsistent in its approach to contaminated sediments. Few aspects of sediment handling, treatment, or containment have been left unregulated, but most applicable laws and rules were not written explicitly to deal with contaminated sediments. As a result, related decisions may not be fully risk-based, and some technically sound management strategies may be foreclosed (a situation that is discussed further in Chapter 3).

The mechanisms of the regulatory process in a given situation depend on where the sediments are located; where they will be placed; the nature and extent of the contamination; and whether the purpose of removing or manipulating the sediment is navigation dredging, environmental cleanup, site development, or waste management (see Appendix B, Table B-1). For example, the dredging of sediment in navigable waters requires a Section 10 permit from the USACE under the Rivers and Harbors Act (RHA). Excavation of sediment from non-navigable waters, or from containment structures, may not be regulated under federal law but could be affected by a variety of state laws. The erection of structures in navigable waters or the emplacement of materials that may obstruct navigation or alter the course, condition, location, or capacity of the waterway may also require a Section 10 permit. This could be the case, for example, where a CDF is constructed to contain dredged material, where dredged material is used to construct an offshore island, or where a clean sand or clay "cap" is used to isolate and contain in situ or deposited sediments.

---

[8] Not all contaminants are dissolved in pore water. PCBs, for example, can be present in the pore space as an organic liquid phase.

Transport for the purpose of dumping dredged sediment in ocean waters (defined as waters beyond the baseline from which the territorial seas are measured) is regulated by the USACE under Section 103 of the MPRSA. Similar discharges in inland or coastal waters on the other side (inland) of the ocean baseline are regulated by the USACE under Section 404 of the CWA, as are discharges of fill material into both inland and ocean waters out to the three-mile limit (the territorial sea). In both cases, affected states may veto or attach conditions to a discharge if it contravenes the state's water quality standards or approved coastal zone management plan. If sediments proposed for ocean disposal are deemed to contain mercury or cadmium compounds, organohalogens, or petroleum hydrocarbons (as other than "trace contaminants") based on prescribed bioassay and bioaccumulation testing procedures, then ocean dumping may be prohibited, although discharge into inland waters may be acceptable as long as the sediments satisfy applicable regulations and guidelines under Section 404 of the CWA. Placement on land is also acceptable, unless the sediments exhibit hazardous waste characteristics (i.e., exceed RCRA regulatory limits, in which case disposal is permitted only at approved RCRA facilities). However, since 1988 the USACE has maintained that dredged materials are not subject to regulation under RCRA, and rule making is pending to clarify this point.

Finally, if contaminated sediments are excavated as part of a remedial response under CERCLA, then they must be treated, contained, or disposed of in a way consistent with applicable or appropriate and relevant regulatory requirements under federal or state law and must meet other Superfund standards. These requirements may impede the cost-effective management of contaminated sediments. Section 121(b) of CERCLA, for example, gives preference to treatments that "permanently" reduce contamination, thereby possibly constraining a site manager's ability to use capping (an issue discussed further in Chapter 5).

A further constraint is imposed on the management of contaminated sediments by the lack of regulatory adherence to the "polluter pays" principle typically followed in other cases of waste management. All too often, point and nonpoint sources of contamination, often far upstream, are not held accountable. As a result, downstream ports seeking to proceed with critical navigation dredging are burdened with extra costs and delays. A situation of this type arose in Newark Bay, a highly industrialized area beset for more than a century by contamination from multiple sources. When the Port Authority of New York and New Jersey applied for dredging permits in 1990, it was required for the first time ever to test for dioxin, which was found to be present at low, part-per-trillion levels. Despite the upstream origin of the contamination, the port had to undertake a series of studies, and, because of interagency disputes over the permit and a lawsuit, the dredging was delayed until 1993 (Weis, 1994).

The fragmented nature of the combined federal and state regulatory framework demands that many parties be involved in decision making, a situation that sometimes results in confusion over who is in charge. Because each regulatory

TABLE 1-1 Time Lapse between Identification of a Problem and Implementation of a Solution: Examples from Six Case Histories[a]

Case Study	Problem Identified	Solution Implemented
Boston Harbor	Problem seen in late 1960s; litigation in early 1980s forced action	1996 or later
Hart and Miller Islands	Permit obtained 1976	Legal challenge resolved 1980; containment structure completed 1984
James River	Fisheries closed 1975	Decision made after 1978
Marathon Battery	Problem seen in early 1970s; NPL listing 1981	1993
Port of Tacoma	Problem seen in 1983	1994
Waukegan Harbor	Problem seen in mid-1970s	1991

[a]These case histories are summarized in Appendix C.

program emphasizes different issues, the lead decision maker may be unsure how to address the related but separate concerns of other agencies and the public, in which case the decision maker may simply request more and more information and analysis or even defer action. The problem is compounded if there is no strong, knowledgeable project proponent who can maintain pressure on the decision maker and keep the regulatory process moving. Even under the best of circumstances, solutions may not be implemented for years.

In the committee's six case histories (summarized in Appendix C), the delay between discovery of a problem and implementation of a solution ranged from approximately 3 to 15 years (see Table 1-1). The problem is not the involvement of many stakeholders but the often adversarial nature of their relationships and the convoluted regulatory path they must follow. The diverse areas of expertise and interests of multiple agencies can be accommodated as long as they are applied in a constructive way to accomplish a logical, risk-based objective.

Political Challenges

The risk posed by contaminated marine sediment is neither easily measured nor highly visible—characteristics that may foster disagreements among stakeholders about how to manage the problem. On land, where contamination may occur in direct proximity to people and food sources or in groundwater that people drink, exposure pathways are clear, and there is a reasonable basis for anticipating sufficient risks to justify a major effort. In the aquatic environment, even when the risk of contaminating the food chain is a real concern, the contamination and exposure pathways are hidden under water and may be difficult to define. In addition, the extent of the threat may be altered by physical and biochemical sequestration mechanisms, which may reduce the bioavailabilty of a contaminant and thereby limit ecosystem effects, including biodegradation.

Regardless of these factors, members of the public and their elected representatives tend to equate the physical presence of contaminants with risk and to insist on more intensive removal and treatment of underwater sediments than of terrestrial contaminants. At the same time, in an era of shrinking federal budgets and dwindling disposal space, it is becoming more important than ever to ensure that management efforts are cost effective. Sometimes conflicts arise between minimizing or eliminating risk and controlling costs. Striking a balance can be a formidable political challenge. Failure to strike a balance among stakeholder interests can delay or stall a project, which was apparent in the committee's case histories (Appendix C). Techniques for meeting the political challenge are discussed in Chapter 3.

Whether motivated by technically sound arguments or emotional self-interest, many stakeholders have common concerns in decisions about managing contaminated sediments. Port communities have powerful economic reasons for dredging. Government regulators are responsible for protecting natural resources and enforcing a complex web of laws and regulations. Environmental groups and community residents who are concerned about public health and natural resource quality are just as committed. They may want remedial action but oppose the deposition of dredged sediment on nearby land or in the ocean.

Management Challenges

Many strategies for managing contaminated sediments are available, some of them very sophisticated. However (as discussed in Chapter 5), many advanced remediation technologies have not been tested extensively at marine sites, and costs can be very high. Superfund cleanup costs can be as high as $1 million per acre (NRC, 1989a). The cost of an entire management plan—dredging, transportation, treatment or containment, and long-term monitoring—must be considered. Costs of these steps vary widely. Dredging is relatively inexpensive per unit volume. At the other end of the spectrum, some treatment technologies have such high unit costs that their use is effectively precluded for treating large volumes of sediment.

Trade-offs often must be made between technology effectiveness and cost. The challenge is to identify the most cost-effective[9] solution for the project at

---

[9] Cost effectiveness is defined here as a measure of tangible benefits for the money spent.

hand and then optimize it by the using systems engineering approaches. When contamination is concentrated in hot spots, effective but expensive treatment options may be feasible. In some instances, it may be cost effective to identify the most highly contaminated sediment and treat the smallest possible volume. In other cases, in order to be acceptable economically and to the public and environmental authorities, large volumes of sediment must be handled, necessitating the use of less costly containment methods. The beneficial reuse of clean or contaminated dredged material can improve prospects for success.

Although there is clearly room to increase the effectiveness and reduce the costs of contaminated sediment management, there is also a built-in bias against innovation. Funding and executing most dredging projects is the responsibility of public agencies, which are subject to the historical constraints on custodians of public funds. These constraints, by design, narrowly focus the contracting process and do not encourage innovative approaches or technologies. In fact, the term "innovative" in this context is often interpreted to mean high risk, an uncertain outcome, and an invitation to post-project censure. Creative management is required to overcome the institutional barriers to innovation. All participants must recognize that an innovative approach may end in failure, and they must agree up front to share the bureaucratic and financial risk.

SUMMARY

The management of contaminated sediments is a difficult problem. The combination of high public expectations, confusing and overlapping jurisdictions, generally low contamination levels, large quantities of affected sediments, risk management challenges, and handling and treatment difficulties may result in large sums of money being spent on partial solutions for low-risk situations. As economic and environmental interests converge and conflict, improved management approaches and technologies need to be developed. Progress in science and engineering have advanced the nation's capability of detecting contaminants; the challenge now is to foster similar advances in decision making and remediation.

There is a conceptual need to balance the risks, costs, and benefits in the face of uncertainties and disagreements about decisions (NRC, 1989b). There is also a practical need to comply with relevant regulations, consider the concerns of all stakeholders, address site-specific considerations, and identify appropriate technologies. Recognizing the multifaceted nature of the problem, the present report is an attempt to set out a risk-based strategy for making management decisions and for selecting remediation technologies. Chapter 2 describes the committee's conceptual management approach, which takes into account the challenges outlined in this chapter.

REFERENCES

Cochran, J.K., D. Hirschberg, and J. Wang. 1993. Chronologies of Contaminant Input to Marine Wetlands Adjacent to Long Island Sound, part 1. Lead-210 and Trace Metals. NOAA Status and Trends Report. Washington, D.C.: National Oceanic and Atmospheric Administration.

Gibbs, R.J. 1973. Mechanisms of trace metal transport in rivers. Science 180:71–73.

Interagency Working Group on the Dredging Process. 1994. The Dredging Process in the United States: An Action Plan for Improvement. Report to the Secretary of Transportation. Washington D.C.: Maritime Administration.

Maritime Administration. 1994. A Report to Congress on the Status of the Public Ports of the United States, 1992–1993. MARAD Office of Ports and Domestic Shipping. Washington, D.C.: U.S. Department of Transportation.

Moore, J.N., E.J. Brook, and C. Johns. 1989. Grain size partitioning of metals in contaminated coarse-grained river flood plain sediments, Clark Fork River, Montana, USA. Environmental Geology and Water Science 14(2):107–115.

Nakles, D.V., and D.G. Linz, eds. In press. Environmentally Acceptable Endpoints in Soil. Annapolis, Maryland: American Academy of Environmental Engineers.

National Research Council (NRC). 1983. Risk Assessment in the Federal Government: Managing the Process. Washington, D.C.: National Academy Press.

NRC. 1989a. Contaminated Marine Sediments: Assessment and Remediation. Washington, D.C.: National Academy Press.

NRC. 1989b. Improving Risk Communication. Washington, D.C.: National Academy Press.

NRC. 1993a. Issues in Risk Assessment. Washington, D.C.: National Academy Press.

NRC. 1993b. Workload Transition: Implication for Individual and Team Performance. Washington, D.C.: National Academy Press.

NRC. 1994a. Science and Judgment in Risk Assessment. Washington, D.C.: National Academy Press.

NRC. 1994b. Building Consensus Through Risk Assessment and Management of the Department of Energy's Environmental Remediation Program. Washington, D.C.: National Academy Press.

NRC. 1994c. Ranking Hazardous-Waste Sites for Remedial Action. Washington, D.C.: National Academy Press.

NRC. 1995. Technical Bases for Yucca Mountain Standards. Washington, D.C.: National Academy Press.

NRC. 1996. Understanding Risk: Informing Decision in a Democratic Society. Washington, D.C.: National Academy Press.

U.S. Army Corps of Engineers (USACE). 1991. Risk Assessment: An Overview of the Process. Environmental Effects of Dredging. Technical Notes, EEDP-06-15. Vicksburg, Mississippi: U.S. Army Engineer Waterways Experiment Station.

USACE. 1995. Continuing Cost Analysis. Unpublished report. Washington, D.C.: USACE Dredging and Navigation Branch.

U.S. Department of Transportation. 1994. Public Port Financing in the United States. Washington, D.C.: Maritime Administration.

Weis, J. 1994. Presentation to the Committee on Contaminated Sediments, National Academy of Sciences held July 13–15, 1994, in Washington, D.C.

2

Making Better Decisions
A Conceptual Management Approach

To meet the challenges identified in Chapter 1, a consistent management approach is needed that systematically takes into account relevant considerations at the proper time. This chapter describes the conceptual basis of an approach that, in the committee's judgment, can be used as the foundation for improved decision making in the development and implementation of effective, comprehensive plans for managing contaminated sediments.

The proposed approach is centered around risk management because contaminated sediments are only a problem to the extent that they pose risks to human health and the environment. The general approach is outlined in the first section of the chapter, which lays out a road map for the development of a management plan in the form of a flow diagram and supporting text. The remainder of the chapter examines the ways risk management comes into play during project planning and implementation. Various perspectives on risk and specific risk-based approaches that can be used to improve decision making are discussed.

CONCEPTUAL VIEW OF THE MANAGEMENT OF CONTAMINATED SEDIMENTS

To provide a framework for a systematic analysis, the committee developed a conceptual overview of the process for managing contaminated sediments (see Figure 2-1). Each element is discussed briefly in this section, and many of the topics are examined in more detail later in this report. It must be emphasized that the diagram appears similar to, but has a different purpose than, the formal decision-making frameworks available for managing contaminated sediments. USACE and the EPA worked together to develop a framework for evaluating alternatives for

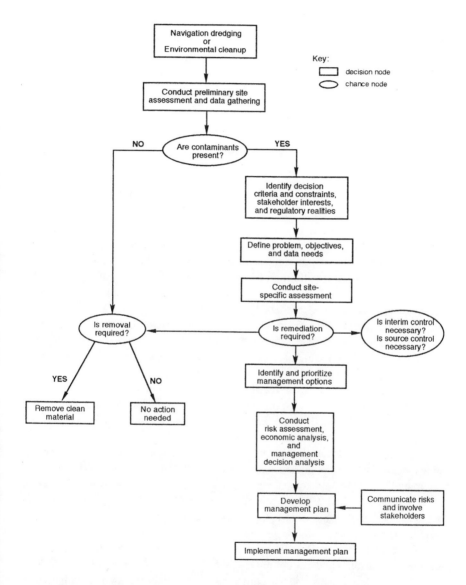

FIGURE 2-1 Conceptual overview of the management of contaminated sediments. Note: For more detail on preliminary site assessment, see Figure 4-1. For more detail on implementing the management plan, see Figure 5-1.

the disposal of dredged material associated with navigation projects (*Code of Federal Regulations*, Title 33, Sections 230 to 250)[1] and for obtaining disposal permits under Section 404 of CWA (EPA, 1994). Another framework was developed for evaluating alternatives for remediation in Superfund projects (EPA, 1994). The committee recognizes the utility of these formal decision-making approaches. Figure 2-1 is intended simply to provide a generic overview of the management process and a context for the various components of the committee's assessment.

The forces that drive an effort to manage contaminated sediments may dictate certain courses of action. As discussed in Chapter 1, the two fundamental driving forces are dredging, which is required to meet port and harbor navigation requirements, and environmental cleanup, which is required to reduce contaminant levels to a specific value.

The preliminary site assessment begins with defining the degree and distribution of contamination, as well as justifying the consideration of taking appropriate actions. The data are used to decide whether and what type of contamination is present, to define the sampling area and density needed to characterize the site more fully, and to identify gaps and uncertainties in the available information that need to be overcome through further surveys, sampling, or studies. This report does not dwell on the initial screening process, focusing instead on how to manage contamination once it has been identified.

If a site is judged to be contaminated, then decision criteria, and the constraints within which actions must be taken, need to be identified. Decision criteria include technical, regulatory, and stakeholder considerations. Technical criteria are related to site characteristics. For instance, some management strategies are more suited to handling small rather than large volumes of sediments; some strategies are appropriate for handling organic contaminants and others for metals; and some are limited to handling sediments with particular physical characteristics, including water content and grain size.

The question of risk also needs to be addressed. What level of risk is acceptable? What levels of contamination are acceptable? Risk can never be eliminated completely, and all management strategies are designed to reduce risk to a certain level at a specific cost. Cleanup standards can be dictated by applicable laws and regulations, which impose a variety of constraints on the management of contaminated sediments. The interests of stakeholders also influence the choice of management strategies. Regulatory realities and stakeholder interests need to be kept in mind prior to and throughout the planning and implementation of a project. Based on the decision criteria and constraints, a problem statement and objectives need to be developed and the need for further data identified. Because it can be

---

[1] References to the *Code of Federal Regulations* will be abbreviated using the format: 33 CFR §230 to §250.

very expensive to fill critical data gaps, refine cleanup standards, and reduce uncertainties in the relevant site parameters, the purposes for which data are to be collected need to be delineated clearly at this stage, and justification needs to be provided for any proposed additional surveys and studies.

The next step is making a detailed site-specific assessment. The volume, distribution, and degree of contamination need to be determined as precisely as necessary and affordable for a specific project. The level of effort required depends heavily on the survey technology and sampling approach. Ideally, the distribution of contamination in three dimensions would be determined so that all unacceptable contamination—and only that amount and no more—could be removed for appropriate treatment. However, in the committee's judgment this is neither technologically nor economically feasible at this time.

Sufficient information needs to be gathered to assess the risks and hazards posed by the contamination, the degree of risk reduction required, and the projected rate of natural recovery if no action is taken. Risk assessment techniques are discussed later in this chapter.

If the level of estimated risk calls for remedial action, then management options need to be identified and organized in order of their applicability. Approaches that cannot be used, because of constraints identified earlier, must be screened out and eliminated, and the remaining strategies ranked according to implementation costs and uncertainties. If the level of risk is very high, then interim controls, such as a ban on fishing, may be implemented immediately. Source control also needs to be considered to eliminate continuing contamination and to ensure that remedial measures will not have to be repeated at a later date.

Once management strategies have been ranked, the most promising options can be compared based on an evaluation of risks, costs, and benefits. In some cases, the best choice is obvious, but in other cases additional data or analytical estimates may be needed. The fundamental issue to be addressed is how best to allocate scarce resources using an integrated set of tools. The latter part of this chapter examines analytical tools—risk analysis, cost-benefit analysis, and decision analysis—for examining trade-offs and arriving at a management plan.

Based on the results of comparisons and evaluations, a comprehensive, long-term management plan needs to be developed that is reasonably certain to meet the remediation criteria and to have the least economic impact, in terms of direct costs and the impact on the local, regional, and national economy. The acceptability of the associated risks and costs is, at this stage, a matter of judgment. The relevant risks need to be communicated effectively to stakeholders, who need to be involved and invested in the decision-making process. The plan needs to be reviewed in light of the mandates of relevant agencies, commercial and business interests, and public concern for the environment and the economy. If the plan is not acceptable, then the objectionable elements need to be removed through reconsideration of the balance of risks, costs, and benefits. If the plan is acceptable, then the associated expenditure of time and money is justified, and the plan can be implemented. The final step is implementation of the management plan.

Systems Approach to Risk-Based Management

To implement a management plan, a systems framework is needed for the engineering feasibility studies, the design, and the optimization of selected remediation technologies. Systems engineering is widely used in the design of complex technological processes to ensure that the various subsystems function together smoothly and achieve optimum overall effectiveness. A formalized approach and discipline are essential for defining potential solutions to the management of contaminated sediments.

Systems analysis, at its simplest, includes the definition of a boundary that surrounds a problem, quantitative representation of how the components within the boundary interface and interact, the constraints imposed on the bounded problem, and an evaluation of alternative ways to meet the agreed goals. The analysis applied in systems engineering can be mathematical, with each goal and constraint specified by exact quantitative algorithms or with model parameters with probability distributions. Fundamentally, however, systems analysis represents a structured approach to developing and improving the design of a complex system and its subsystems, in light of overall project goals and objectives. The approach, which is discussed further in Chapter 5, involves trade-off studies addressing design alternatives, technical and operational considerations, system performance, risks, costs, and benefits.

Finally, on completion of the steps identified in the management plan, the residual risks at the site need to be assessed to ensure that the goals have been met. If the residual risks are unacceptably high, then an iteration of the decision-making process may be necessary. It also can be useful to examine whether predictions made during the decision-making process proved to be accurate so as to inform future decision-making processes related to other contaminated sediment sites.

TRADE-OFFS IN RISKS, COSTS, AND BENEFITS

A central feature of the risk-based management approach within a systems framework is the delineation of trade-offs in risks, costs, and benefits that need to be made in choosing the best course of action among available alternatives. The fundamental difficulty involved in making decisions about how best to manage contaminated sediments lies in the measurement of the gains and losses to various stakeholders. For example, the well-being of ports and the general public is advanced by dredging, but the benefits must be weighed against the risks and costs of dredging and managing sediment. Similarly, the benefits of environmental cleanup to human health and ecosystems must be weighed against the costs. The chances of identifying and implementing the best possible solution are enhanced when stakeholders have a solid understanding of all the gains and losses associated with various alternatives.

A number of decision-making tools can be used to determine trade-offs in risks, costs, and benefits. These tools include risk analysis, cost-benefit analysis, and decision analysis. Risk analysis involves the extended application of risk assessment techniques, which typically are used only to assess the severity of in-place contamination. Cost-benefit analysis examines the costs associated with the reduction in risk to acceptable levels as established by risk assessments. Decision analysis incorporates the data from cost-benefit analysis into a computational framework that estimates the outcome of selected management approaches and evaluates the relative merits of alternative courses of action.

Risk Analysis

Risk analysis encompasses risk assessment and risk management, concepts defined in Chapter 1, as well as risk communication (USACE, 1991; EPA, 1996, and references therein). Risk communication is a dialogue that takes place on two levels, first when the risk assessor communicates technical findings to the risk manager, and later when the risk manager conveys the results to the public and other stakeholders (see Chapter 3). All three aspects of risk analysis—assessment, management, and communication—are essential to the cost-effective management of problems in general (NRC, 1996) and to the management of contaminated sediments in particular. Currently, however, they are not incorporated into the contaminated sediment management process. In fact, all three aspects of risk analysis are seldom included in any single project.

As noted in Chapter 1, risk assessment typically is used only to determine the hazard posed by the initial contamination. Various methods, some more rational than others, have been used to conduct risk assessments. After the initial risk assessment, however, risk may not be addressed again directly in the sediment management process. There is little direct regard for the risks associated with sediment removal or relocation or for post-project residual risks. Although risk reduction capability is a consideration in the selection and evaluation of sediment management strategies, this capability typically is only predicted or estimated, not measured. For example, the efficacy of remediation technologies is usually monitored by measuring physicochemical parameters rather than by assessing residual risks (see Chapter 5). The absence of quantitative data on risk reduction capabilities complicates attempts to evaluate strategies for the disposal, remediation, and beneficial use of sediments.

But the fundamental reason for not using risk analysis more lies in the uncertainties inherent in current risk assessment techniques,[2] which are subject to debate and have several limitations: They provide only approximations of effects on human health and the environment; they provide evidence of acute, not chronic (such as on reproduction or growth), effects; and they do not take into account all conditions at a test site. Without a quantitative link between accepted measures of sediment quality and corresponding risks to the ecosystem and human beings, there will continue to be disagreements concerning the magnitude of the original problem and the efficacy of various remediation strategies (see Box 2-1). Although the resolution of these issues is outside the scope of the present report, it seems clear that improved end-points must be developed and interpreted constructively. In the meantime, however, decisions need to be made, and risk analysis can be used to improve these decisions despite the inherent uncertainties.

Improved techniques for measuring risk would help conserve scarce resources by ensuring that money is not wasted on unnecessary remediation and by providing end-points for the quantitative evaluation of strategies for the disposal, remediation, and beneficial use of sediments.

Thus, there are significant opportunities for improving and extending the application of risk analysis in the contaminated sediment context. The importance of risk analysis reaches beyond the issues just discussed because the results of risk assessments are essential elements in the cost-benefit analysis and the decision analysis.

Cost-Benefit Analysis

To make decisions about contaminated sediments, decision makers need to weigh the relevant factors, including costs and benefits, and make trade-offs. Risk assessment can provide information about the exposure, toxicity, and other aspects of the contamination, but relying on this approach alone can result in the less-than-optimum allocation of resources unless additional information is considered. For example, even though the concentration of contaminants at a particular site could be toxic enough to induce mortality in a test species, this fact, by

---

[2] The risk assessment paradigm applied by the EPA to human health issues includes source and release assessment (hazard identification), exposure assessment, dose-response (or effects) assessment, and risk characterization. When this fundamental paradigm was reviewed and reevaluated for applicability and efficacy from the standpoint of ecological assessment, a fifth step was needed: consideration of simultaneous or alternate potential sources of environmental perturbation. The EPA framework for ecological risk assessment (see EPA, 1996) includes the following steps: problem formulation (planning, site characterization, stressor characterization, end-point characterization); analysis (exposure assessment, effects assessment); and risk characterization (exposure and effects comparison, determination of uncertainty and limitations, evaluation of ecological significance).

itself, does not indicate the spending level that would be justified for cleanup. The decision must be a determination of the most efficient way to allocate resources based on information, risks, and costs.

Cost-benefit analysis, which makes use of information provided in risk assessments, is a widely used tool that can provide a comprehensive understanding of the trade-offs implicit in choices among dredging or disposal alternatives. (Costs and benefits are defined more completely in Appendix D.) Benefits are the public's willingness to pay for all aspects of the project. Costs are the "opportunity costs," including all the production factors used in construction of the project. Benefits include direct services, such as transportation, as well as indirect services, such as the value of ecological protection.

The nature of the choice is illustrated in Figure 2-2. For purposes of this example, the objective is to relate the amount of contaminant removed from the sediment to the costs and benefits associated with its removal. It is assumed that the magnitude of costs and benefits related to various dredging strategies are known. The vertical axis measures the costs and benefits of removing contaminants; the horizontal axis measures the percentage removed. As the percentage of contaminants removed increases, the costs increase at an escalating rate because it becomes more and more difficult to locate and eliminate the remaining contaminants. At the same time, the benefits of contaminant removal accrue at a decreasing rate, so that additional removal continues to provide benefits but in smaller and smaller increments. The best decision is point A, at which the difference between costs and benefits is the greatest. A poor decision would be point C, at which the costs are greater than the benefits. At point B, the benefits just offset the costs. The important thing is that there are trade-offs associated with every course of action, regardless of the approach used to select that alternative.

Many federal agencies use cost-benefit analyses extensively and have guidelines that explain how costs and benefits are to be computed and used (Water Resources Council, 1983). These concepts can be readily applied to decisions about environmental issues but have not been used systematically in the contaminated sediments context, except when new-construction dredging is involved, in which case cost-benefit analysis is required.

Cost-benefit analyses can be useful for evaluating proposed management strategies. The basic principle is that activities should be pursued as long as the overall gain to society, correctly measured, exceeds the social cost. The difficulty lies in measuring the benefits and costs, or, more to the point, in projecting what they will be before a strategy is implemented. The method of computing cost-benefit ratios depends, in part, on how costs and benefits are defined. Three types of costs are involved in contaminated sediments cases: dollar costs of remediation and cleanup, dollar costs of foregone port services, and environmental costs. None of these costs can be measured precisely. The benefits of an action are simply the costs of not taking that action. Uncertainties concerning the costs of remediation, a major focus of this report, are addressed in Chapter 5. The difficulties involved

> **BOX 2-1**
> **Evaluating Sediment Contamination: Effects-Based Testing and Sediment Quality Criteria**
>
> Three specific situations or reasons exist for evaluating sediments (Brannon and McFarland, 1996). The first is to determine what unacceptable adverse effects, if any, navigation channel sediments will pose in a particular placement environment. EPA regulations 40 CFR §220 to §228 and 40 CFR §230 provide guidance on the aquatic placement of dredged material. The second reason is to determine what effects on aquatic ecosystems sediments may have if they are left undisturbed or if they are removed for environmental purposes. If sediments are determined to have unacceptable environmental effects, consideration may then be given to some type of remediation, which may or may not include removal. If sediments are to be removed, then the potential effects they will have at the placement site must be considered. The third reason, recently advanced by the EPA, is for the source control of contaminants. Determination of locations where sediments, as sinks for contaminants, have unacceptable environmental or human health impacts could lead to identification of the source of the contamination.
>
> Effects-based testing is currently the primary means of sediment evaluation and is a basic tool for estimating the risk of various sediment management techniques (dredging, cleanup, etc.) to the aquatic environment. The assessment of sediment quality is a hazard assessment intended to determine whether the exposure of aquatic biota to a sediment will cause an increase in the incidence of adverse, unacceptable effects. To supplement effects-based testing, the EPA also is developing sediment quality criteria (SQC) as a way of determining the potential biological impacts of contaminants in sediments (DiToro et al., 1991) and has published proposed numerical SQC in the *Federal Register* for public comment (*Federal Register*, vol. 59, no. 11, January 18, 1994, p. 2652).
>
> Effects-based testing involves the use of organisms to determine the biological effects of sediments. In general, test species are exposed in the laboratory to sediments being evaluated, and their response is compared with that of reference organisms with regard to specific biological end-points, such as mortality. Effects-based testing inherently accounts for all of the contaminants present in a sediment and the potential interactions among contaminants because the approach relies on the exposure of test species to whole sediment. Therefore, the precise chemical composition of the sediment need not be known. Potential interactive effects of multiple contaminants are integrated based on the response of the test organism.
>
> Much of the criticism of the effects-based approach for evaluating dredged material centers on the lack of chronic, sublethal test end-points

> (i.e., growth and reproduction) in the current regulatory program. However, several chronic, sublethal sediment toxicity tests are in the late stages of development (Dillon et al., 1995; Emery and Moore 1996; Liber et al., 1996) and may now be used as part of a dredged material evaluation. In addition, bioaccumulation tests account for the uptake of contaminants over longer-term exposures (28 days, and if necessary steady state can be estimated) and may be used to infer the potential for chronic, sublethal effects.
>
> The EPA is developing SQC pursuant to the CWA, §304(a)(1) and §118(c)(7)(c), which are aimed at protecting benthic organisms from chronic sediment toxicity. The SQC approach advocated by the EPA for estimating the potential risk posed by contaminated sediment uses equilibrium partitioning modeling to predict pore water concentrations of nonpolar organic compounds. These predicted pore water concentrations are then compared to chronic water quality criteria as an effects threshold. The EPA has proposed that SQC be used both in preventing sediment contamination and in establishing cleanup targets.
>
> SQC are single contaminant criteria, yet sediments typically contain a complex mixture of contaminants. Regulatory assessments, such as dredged material evaluations, require that the interactive effects of sediment contaminants be evaluated. Sediments are likely to contain contaminants for which SQC do not exist, which means that effects-based testing will still be required to determine whether exposure of aquatic biota to a sediment will cause an increase in the incidence of adverse, unacceptable effects.
>
> Government agencies embrace effects-based testing as a basis for making decisions concerning the placement of sediments and are moving to develop chronic effects-based testing protocols and applying more formalized risk assessment to bioaccumulation test results. Further development of chronic tests will provide improved end-points. There would still be a need to understand and interpret those end-points in a regulatory context to determine what constitutes an unacceptable adverse effect.

in quantifying the value (costs) of economic services provided by ports and the environmental costs of remediating (or not remediating) contaminated sediments are discussed in Appendix D.

Although the measurement of costs and benefits can be laborious, it is worth the effort in projects where the stakes are very large. Even for small projects for which detailed measurements may seem impractical, a consideration of economic issues can be useful for making qualitative judgments about management strategies. For example, remediation technologies can be evaluated on the basis of cost

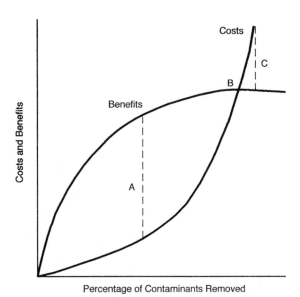

FIGURE 2-2 Conceptual illustration of the trade-offs involved in cost-benefit analysis. A = best decision point; B = benefits equal costs; C = worst decision point.

effectiveness, at least in a qualitative sense (quantitative comparisons are precluded because of insufficient data on both cost and effectiveness [see Chapter 5]). Also, basic economic principles suggest some guidelines for decision making in general. (The derivation of these guidelines is explained in Appendix D.) For example, certain initial measures to reduce contamination may be relatively inexpensive, whereas the corresponding social returns can be quite high. However, the extensive cleanup of contaminated sediments tends to become increasingly costly as the concentration of contaminants declines. Furthermore, the social gains from cleanup tend to increase more slowly as the contaminant concentration declines. Decision makers are cautioned against seeking extreme solutions without first measuring social benefits and costs.

The appropriate use of cost-benefit analyses could be encouraged through changes in federal policies and practices. For example, although cost-benefit analysis is currently required only for new-construction dredging, it might also improve decision making in situations that require major or continuing maintenance dredging where, even though the need to dredge has been established, decisions still need to be made concerning many other variables. Unfortunately, USACE guidelines for cost-benefit analyses are not complete and are not followed in all cases. The guidelines discuss, for example, how to account for situations in which additional traffic is encouraged at one port at the expense of another port, but these provisions are not employed in practice. The guidelines do not even address the possibility of price changes in navigation services as a result of changes in national policies. Although a detailed analysis of these issues is outside the scope of this report, it is clear that efforts to improve the precision, completeness, and ease of use of cost-benefit analyses could improve decision making.

Decision Analysis

Decision makers need to know how to use information about risks, costs, and benefits that may be controversial and difficult to evaluate, compare, or reconcile. The committee devoted considerable attention to finding ways to meet this need, which was identified in an earlier NRC report (NRC, 1989). Indeed, the demand is becoming increasingly urgent as the number of proposals for the remediation of contaminated sediments grows and as costs and controversies multiply.

One tool that can help resolve problems with many variables is decision analysis, a computational technique for estimating the outcomes of management approaches. Decision analysis does not provide absolute solutions but can offer valuable insights. It can integrate the results of key management tasks (e.g., risk assessment, site assessment, economic assessment, technical feasibility studies) into models of the problem as it appears from the perspectives of various stakeholders. The modeling approach allows stakeholders to explore disagreements about subjective elements of the problem, thereby expediting problem solving. The process also formally accounts for uncertainties. The output is the identification of the optimum approach, that is, the strategy that offers the best odds for successful risk management.

Although decision analysis is not a new technique, it apparently has yet to be used in managing contaminated sediments. The committee's assessment, including its application of decision analysis to a hypothetical test case involving remediation of a hot-spot contamination site (see Appendix E), suggests that this approach may be valuable for sorting out management options when more conventional methods fail. Decision analysis is applicable in highly complex, contentious situations because it can accommodate more variables (including uncertainty) and perspectives than other analyses, such as cost-benefit analysis, that measures a single outcome, and because the methodology of decision analysis is explicit and rigorous and the analytical pathways are reproducible. However, because decision analysis is technical in design and involves complex computations, it will take some time and effort for stakeholders to gain confidence in the approach.

Remediation of contaminated sediments tends to be expensive and arduous, so any approach that helps expedite corrective action and resolves environmental controversies fairly and cost effectively could prove valuable. Decision analysis appears to be such an approach. Its use may be particularly timely now because recent advances in computer hardware and software now make it possible to perform user-friendly, interactive analyses.

SUMMARY

The risk-based approach outlined in this chapter provides a rational strategy for the cost-effective management of contaminated sediments. This discussion also highlighted opportunities for improving management through the use of risk analysis, cost-benefit analysis, and decision analysis.

Risk analysis is essential to the cost-effective management of contaminated sediments but has not been applied as often as it might be. At present, risks typically are assessed only for the initial contamination, and little or no consideration is given to the risks of sediment removal or relocation or the risks remaining after remediation. This approach limits the capabilities of evaluating strategies for sediment disposal or remediation, and opportunities for the beneficial use of sediments. The scientific underpinning of risk analysis also requires attention. Various types of sediment quality criteria are under development, but these approaches have not been linked quantitatively to ecological or human health risks. Environmentally acceptable end-points are needed for sediment contamination.

In the contaminated sediments context, cost-benefit analysis usually is used only for major new navigation dredging projects and tends to be narrow in scope. The use of cost-benefit analysis could be extended to help identify the optimum solution for managing contaminated sediments. From an economic standpoint, the best strategy is the one in which benefits outweigh the costs by as much as possible. The costs involved are difficult to calculate and uncertain, but comprehensive cost-benefit analysis can still be worth the effort in very expensive or extensive projects. Informal estimates or cost-effectiveness analysis may suffice in smaller projects. There is also room for improvement in federal guidelines for the computation and use of benefit and cost data. For example, the guidelines do not take into account the economic effects of shifts in transportation patterns or changes in the prices of navigation services.

Decision analysis offers a way to balance the consideration of risks, costs, and benefits of various strategies for managing contaminated sediments. Decision analysis could be particularly valuable because it can accommodate more variables (including uncertainty) and different perspectives than other techniques, such as cost-benefit analysis, that measure single outcomes. Decision analysis also can serve as a consensus-building tool by enabling stakeholders to explore the subjective elements of problems and, perhaps, find common ground. However, because it is technical in design and involves complex, albeit logical, computations, decision analysis is probably worth the effort only in exceptionally complicated and contentious situations in which stakeholders are willing to devote the time to gain confidence in the approach.

REFERENCES

Brannon, J.M., and V.A. McFarland. 1996. Technical Considerations for Sediment Quality Criteria. Paper presented at Water Quality '96. Environmental Engineering and Ecosystem Management, U.S. Army Corps of Engineers, 11th Seminar on Water Quality held February 26, 1996–March 1, 1996, in Seattle, Washington.

Dillon, T.M., D.W. Moore, and D.J. Reish. 1995. A 28 day sediment bioassay with the marine polychaete, *Nereis (Neanthes) Arenaceodentata*. Pp. 201–215 in Environmental Toxicology and Risk Assessment, vol. 3. J.S. Hughes, G.R. Biddinger, and E. Mones, eds. ASTM STP 1218. Philadelphia: American Society for Testing and Materials.

DiToro, D.M., C.S. Zarba, D.J. Hansen, W.J. Berry, R.C. Swartz, C.E. Cowan, S.P. Pavlou, H.E. Allen, N.A. Thomas, and P.R. Paquin. 1991. Technical basis for establishing sediment quality criteria for nonionic chemicals using equilibrium partitioning. Environmental Toxicology and Chemistry 10:1541–1583.

Emery, V.L., and D.W. Moore. 1996. Preliminary protocol for conducting 28-day chronic sublethal sediment bioassays under the estuarine amphipod *Leptocheirus plumulosus* (Shoemaker). In Environmental Effects of Dredging. Technical Note EEDP-01-36. Vicksburg, Mississippi: U.S. Army Engineer Waterways Experiment Station.

Environmental Protection Agency (EPA). 1992. An SAB Report: Review of Sediment Quality Criteria Development Methodology for Non-ionic Organic Contaminants. EPA-SAB-EPEC-93-002. Washington D.C.: EPA.

EPA. 1994. Assessment and Remediation of Contaminated Sediments (ARCS) Program. Remediation Guidance Document. Great Lakes National Program Office. EPA 905-R94-003. Chicago: EPA.

EPA. 1996. Draft Proposed Guidelines for Ecological Risk Assessment. Risk Assessment Forum. Washington, D.C.: EPA.

Liber, K., D.J. Call, T.D. Dawson, F.W. Whiteman, and T.M. Dillon. 1996. Effects of *Chironomus tentans* larval growth retardation on adult emergence and ovipositing success: Implications for interpreting freshwater sediment bioassays. Hydrobiologia 323:1–13.

National Research Council (NRC). 1989. Contaminated Marine Sediments: Assessment and Remediation. Marine Board. Washington, D.C.: National Academy Press.

NRC. 1996. Understanding Risk: Informing Decision in a Democratic Society. Washington, D.C.: National Academy Press.

U.S. Army Corps of Engineers (USACE). 1991. Risk Assessment: An Overview of the Process. Environmental Effects of Dredging. Technical Notes EEDP-06-15. Vicksburg, Mississippi: U.S. Army Engineer Waterways Experiment Station.

Water Resources Council. 1983. Economic and Environmental Principles and Guidelines for Water and Related Land Resources Implementation Studies. Washington, D.C.: U.S. Government Printing Office.

3

Forces Influencing Decision Making

A strategy for managing contaminated sediments needs to address the four challenges outlined in Chapter 1 within the general management process presented in Chapter 2. The committee's assessment of the best way to meet the challenges is organized into four general thematic areas, which provide the structure for the remainder of this report. All four themes require attention in the development of an effective management strategy:

- regulatory realities
- stakeholder interests
- site-specific considerations
- remediation technologies

These four themes respond generally to the challenges outlined in Chapter 1 (regulatory and legal, political, chemical, and management and technological), although there is not always a one-to-one relationship.[1] The themes also come into play repeatedly in the conceptual management process outlined in Chapter 2.

Organization of the report around these four themes was considered necessary for the committee to make a coherent analysis and respond most directly to the statement of task. This chapter examines the first two themes—regulatory realities and stakeholder interests—external forces that are sometimes influential

---

[1] The regulatory realities theme encompasses both regulatory and legal challenges, whereas the stakeholder interests theme corresponds to political challenges. The last two theme areas are both concerned with chemical challenges. All four themes address management and technological challenges.

enough to predetermine or strongly shape decisions. The decision maker has greater leeway in dealing with site-specific considerations (see Chapter 4) and remediation technologies (see Chapter 5).

It is apparent that each of the four considerations narrows the choice of management strategies and that failure to consider any of them could undermine the effectiveness of a management plan. In combination, these themes address key aspects of the committee's charge, particularly the tasks of assessing the best management practices and addressing how information about risks, costs, and benefits can be used to guide decision making. The remainder of this report is devoted to an examination of the four essential considerations and an analysis of the issues requiring attention and the opportunities for formal changes that could facilitate the management of contaminated sediments in general. The analysis encompasses many of the lessons learned from the case histories (summarized in Appendix C).

The first consideration, regulatory realities, is paramount. The regulatory framework dictates many of the choices facing decision makers, and attention to its nuances can save time and money. As summarized in Chapter 1 and addressed more thoroughly in Appendix B, a confusing array of federal and state statutes govern, and often impede, decision making in contaminated sediment management. For the project proponent to achieve the project objective (e.g., dredging a channel, cleaning up a contaminated body of water), regulatory requirements and constraints must be fully factored into the decision-making process. In some cases, legislative constraints may frustrate the achievement of an optimum balance among risks, costs, and benefits.

Consideration of competing stakeholder interests is key to the timely implementation of solutions, which can be delayed for years or even decades if major disputes arise (see Table 1-1). Although many decisions associated with the management of contaminated sediment are driven purely by engineering and fiscal considerations, other aspects of the process are value driven. Remediation endpoints, balancing of various risks, and political acceptability are among the more notable value-driven components of the management process. But these values are rarely absolute. Therefore, it is essential that project proponents involve stakeholders early in the decision-making process to ensure that various viewpoints and concerns can be clarified and consensus building can begin.

Project-specific considerations include information-gathering and engineering related to the site in question. These must be handled properly for management efforts to be successful. Project-specific considerations include source control, site characterization, and characterization of the nature and extent of contamination. The key challenge is to determine the types and levels of analysis required—that is, to identify the amount of information and engineering that is both necessary and feasible to support site-specific judgments.

Further constraints on possible solutions are imposed by the state of the art in remediation technologies. Given unlimited time and money, any contaminated sediment site could probably be cleaned up using technologies currently available. But resources are always limited, and treatment technologies—the most effective solutions for eliminating contaminants rather than simply containing them—evolve slowly with time and are very expensive. To make the best of a less-than-ideal situation, therefore, the decision maker must select appropriate, available, and affordable technologies and optimize their effectiveness as part of an overall remediation system or process.

REGULATORY REALITIES

In examining the regulatory framework from the perspective of the committee's task (which calls for an assessment of best management practices and an examination of how information about risks, costs, and benefits can be used to guide decision making), the committee determined that the current regulatory scheme does not always promote efficient, cost-effective solutions to problems of contaminated sediments. A fundamental flaw is the apparent inability of regulatory agencies to implement mandated procedures designed to ensure that management decisions reflect an appropriate balance of risks, costs, and benefits. (This is a shortcoming of governing statutes and regulations, not a criticism of the regulatory agencies charged with their implementation.) Although the committee focused primarily on scientific and technical issues, it also recognized that the best management approaches cannot be implemented without a supportive regulatory framework. This section examines key limitations of that framework with respect to the evaluation of disposal alternatives, the timeliness of decision making, cost allocation, and the shortage of placement options.

The ability of agencies to translate sometimes highly technical data into sediment management decisions is not fostered by the regulatory process. As the capabilities for detecting chemical contaminants increase, the level of detail can also be expected to rise. Unfortunately, understanding how the information can best be used in decision making has not kept pace with advances in science, which will make future decision making even more difficult.

Evaluation of Placement Alternatives

A lack of coherence is evident in current procedures for evaluating placement and management alternatives. The regulatory process under MPRSA places primary emphasis on the intrinsic toxicity of the constituents of dredged material. The process involves biological testing of dredged material to determine if the material proposed for dumping beyond the baseline of the territorial sea will cause unreasonable degradation or endangerment of the marine environment or human health. However, the procedures do not consider fully site-specific conditions (e.g., proximity to shellfish beds, other sensitive receptors, food-chain carriers, or the containment of contaminants by an engineered clean cap) that may influence the impact on various organisms.[2] An even more rigid approach is used under the London Convention of 1972, which the MPRSA has made binding under U.S. law. This international treaty categorized the acceptability of materials proposed for ocean dumping on the basis of whether the materials contain certain enumerated "black list" (Annex 1) constituents "other than trace contaminants" or "gray list" (Annex II) constituents that require "special care." Although the rigid MPRSA criteria (i.e., focusing on intrinsic toxicity) may be the safest overall approach from the standpoint of environmental protection, such rigidity can obstruct efforts to reach the best decision in a particular case and can result in the needless waste of resources (e.g., requiring placement on land rather than less costly ocean dumping even when there is no risk-based rationale for prohibiting ocean dumping).

The counterpart procedures under the CWA consider chemical and physical as well as biological characteristics in assessing whether the discharge of dredged material will cause unacceptable adverse impacts. Although far from a risk-based process, these procedures at least do not specify absolute pass-fail criteria, and they are geared to the identification of the least environmentally damaging alternative that is also practical.

Risk reduction is emphasized under the Superfund remedial action program. Site-specific remedies are chosen based on "exposure assessments" during the feasibility study, and remedial alternatives are identified based on their capability to reduce risks of exposure to an acceptable level. But there are no risk-based cleanup standards for underwater sediments at present. The final selection now involves choosing the most cost-effective alternative.

In sum, each set of regulations uses a different approach to assess remedial alternatives, and none considers fully the risk posed by contaminated marine sediments. Although inconsistency alone is not necessarily a major problem, when it is coupled with insufficient attention to risk, it can impede the cost-effective management of contaminated sediments. Cost effectiveness is further impeded by the failure of the MPRSA and Superfund to consider fully the practicality of remedial alternatives, including their economic and technological viability. The CWA does take these issues into account, although, perhaps, it does not emphasize them sufficiently.[3] Risks need to be considered more fully to ensure that they are not underestimated or overestimated

Similar inconsistencies and inattention to risk are evident in the permitting processes for sediment placement. It is currently necessary to secure different

---

[2] As discussed in Chapter 2, risk is a function of both the inherent toxicity of a material (which is evaluated by the MPRSA's biological testing procedures) and the potential of organisms in the receiving environment to be exposed to that material (a factor not directly considered).

[3] Under the CWA, practicability is defined as "available and capable of being done after taking into consideration cost, existing technology, and logistics in light of overall project purposes" (40 CFR §230.3[q] and §230.10[a][2]).

types of permits to (1) place sediments in navigation channels or in ocean waters as part of the construction of land or containment facilities (under the RHA and CWA), (2) dump sediments in the open ocean (under the MPRSA and RHA), (3) dispose of sediments in inland waters or wetlands (CWA), and (4) contain contaminated sediments on land where there is no runoff into waters of the United States (RCRA). The regulations also distinguish between sediments removed during navigation dredging (CWA or MPRSA) and sediments excavated for environmental remediation (Superfund). The EPA and USACE, to their credit, have made substantial progress under existing law toward developing and applying parallel procedures in evaluating dredged material under the MPRSA and CWA (EPA and USACE, 1991, 1994), and the EPA has proposed a national strategy for managing contaminated sediments to promote greater consistency in the evaluation and regulation of contaminated sediments with other EPA programs (EPA, 1994).[4] However, the diverse statutes under which the EPA and USACE operate often impose different constraints on the ability of regulators to balance overall risks, costs, and benefits.

As an example of the anomalies created, the current regulatory regime does not adequately address risk management, focusing instead on the type of activity—removal, containment, or treatment. This misdirected attention can lead to wasted time, energy, and expense, not to mention the possible failure to reduce the risks to human health and the environment.

The problem could be overcome, in part, by the development of a consistent or parallel set of risk-based regulatory requirements for evaluating dredged sediments that do not differentiate (absent a compelling technical justification) among inland, estuarine, and ocean placement but do take into account site-specific biological, chemical, and physical conditions that bear on risks to the environment and human health. To be complete, the regulatory scheme must also consider the relationship between environmental and economic costs and benefits. The overall effect of these changes would be to raise the regulatory focus from mechanical details (i.e., the type and location of dredging and disposal) to a higher level: the cost-effective management of risks to human health and the environment.

Timeliness of Decision Making

Another problem posed by the current regulatory framework is the potential for unnecessary delays. Timely decision making is important to minimizing costs, given that delays can impose both economic and environmental costs. Federal statutes (e.g., the CWA and MPRSA) that involve more than one agency make it difficult, if not impossible, to reform regulatory procedures to facilitate the

---

[4] EPA's draft document describing a strategy for the management of contaminated sediments has been made available for public comment.

administrative process. The regulations, which have evolved over many years, have led to widely varying time frames for decision making. Superfund remedial response can often take many years to identify and implement. Permit decisions under the MPRSA and CWA can be made within a few months of the application when small quantities of uncontaminated dredged material are involved and placement alternatives are identified, but decision making can take many years for larger navigation dredging projects or for projects with complex problems of sediment placement.

The Port of New York, for example, recently had to wait more than three years to obtain a maintenance dredging permit. In the meantime, for various reasons, such as increasing technical demands, the cost of the dredging project rose from $1 million to $17 million (Maritime Administration, 1994). One problem is the multiplicity of federal and state regulatory and resource agencies involved in the management of dredged sediments. Another problem is that the USACE, as lead agency, is confronted with difficult decisions when the placement of contaminated sediment is controversial and stakeholders have been unable to arrive at a clear consensus. A federal interagency working group, under the auspices of the Maritime Administration, recently recommended a series of steps to improve the timeliness of decisions concerning dredging permits (Interagency Working Group on the Dredging Process, 1994). A National Dredging Team composed of seven federal agencies was formed to implement the recommendations of the 1994 interagency report and to serve as a forum for resolving dredging issues. Regulatory reform initiatives pending in the U.S. Congress range from broad-based reforms to specific reforms to realign or consolidate permitting authority under the CWA.

Timely decision making can be facilitated by interpreting regulations based on the *intent* of the underlying statute(s). The EPA has shown a willingness, on occasion, to be flexible, within legal constraints, in the application of regulations. This was demonstrated in the Port of Tacoma case history, where CWA restrictions on avoidable discharges of dredged material were interpreted in such a way that implementation of an innovative (and ultimately successful) cleanup plan was permitted.[5] This enabled the port to implement a creative solution, which simultaneously enabled a stalled navigation dredging project to move forward, an

---

[5] Section 404 of the CWA prohibits the deposition of dredged materials in wetlands, mudflats, or other "special aquatic sites" unless there is no practical alternative judged to be less environmentally damaging. In the Port of Tacoma case, the EPA expressed a preference for near-shore disposal only "in conjunction with projects that otherwise would be permitted as commercial development" (which otherwise would be permissible and would require separate fill projects). This approach enabled the EPA to approve the Tacoma project as one that minimized physical impacts to the near-shore environment by averting filling solely for the purpose of sediment disposal. In addition, while focusing on the objectives and intent of the CWA and Superfund, the EPA and USACE were flexible in interpreting the decision-making criteria, determining that there were no "more environmentally suitable" (as opposed to "less environmentally damaging") alternatives.

intractable Superfund problem to be resolved, and some prime waterfront acreage to be gained by the port. But this commendable focus on underlying objectives is discouraged rather than promoted by the current regulatory framework, which, as demonstrated by the examples given earlier in this section, tends to specify rigid criteria and procedures. One way to promote the achievement of objectives is to emphasize risk-based end-points rather than specific processes. The development of site-specific, environmentally acceptable end-points may provide risk-based performance standards. In the meantime, flexible interpretations of the regulations may be helpful.

Cost Allocation

Cost allocation is another area in which regulations may hinder efficient decision making. One potentially counterproductive situation is created under the WRDA (Water Resources Development Act) of 1986 (§101 and §102 [P.L. 99–662]), which requires local sponsors of federal navigation projects to bear full responsibility for the construction, operation, and maintenance of dredged material disposal sites on land. This provision creates a bifurcated approach to cost sharing, in that the federal government pays for most (usually 75 percent[6] of) new-work dredging and, with the help of a trust fund that collects user fees, all maintenance dredging. But the government does not pay for the costs of sediment placement on land. The project cooperation agreement also creates two unfortunate incentives. First, because the project sponsor must pay for disposal on land (whereas placement in open water is "free") a strong preference is created for placement in open water, whether or not it is in society's (or the environment's) best interest. Second, an approach that places the full cost of land-based placement on the project sponsor creates little incentive for the sponsor to seek out opportunities for the beneficial use of sediment, which usually add to the project cost and may benefit a party other than the proponent. (Beneficial uses are discussed later in this chapter.)

Cost allocations for dredging are also inconsistent. Ports are not required to share directly in the costs of maintenance dredging, but federal requirements under WRDA 1986 compel local sponsors to share in the costs of new-work dredging, with the percentage depending on channel depth. This distinction between the two types of dredging may not be justified economically. Although this complex issue exceeds the scope of the committee's analysis, it could be addressed separately. In any case, problems with the cost allocation scheme for sediment placement must be addressed. To ensure that decisions are not distorted by ill-

---

[6] Costs of new-work dredging are shared by local sponsors and the federal government, and the cost-sharing percentage is based on channel depth. In most cases, the federal share ends up being 75 percent.

conceived or unjustified cost burdens, it may make sense to develop equitable cost-sharing formulas for all the dredging and disposal elements of federal projects. At the same time, it would be helpful if consistent approaches to cost-benefit analysis were applied. Currently, an elaborate system of weighing costs and benefits must be used for new-work dredging. Cost-benefit analysis for maintenance dredging is applied inconsistently, and alternatives for the placement of dredged material are initially based on compliance with environmental regulations in which cost is one factor in decision making (see Chapter 2). Issues of cost need to be addressed systematically because an inconsistent or incomplete consideration of costs can encourage an irrational allocation of scarce resources.

Shortage of Placement Space

Even if the changes outlined above were made, there would still be the problem of limited placement space for contaminated dredged materials, an issue that defies easy answers.[7] Large coastal ports, as well as owners of marine terminals and small private berths, are finding it increasingly difficult to find space for the placement of sediments unsuitable for open-water disposal. Although the development of risk-based strategies for regulating the placement of contaminants in dredged material may reduce the quantity of material requiring land-based management, local ports and other private dredging proponents will always be faced with a shortage of placement sites on land. Constraints include dwindling open space, the logistics of transportation and other handling issues, and public opposition to the placement of contaminated materials near populated areas. Whatever the reasons, capacity shortfalls could limit dredging and have significant negative socioeconomic consequences at the local, regional, and even national level.

Given the national interest in achieving and maintaining adequate dredging depths at certain key ports, it may be counterproductive to place the entire burden of finding and funding land-based placement sites for dredged material on local interests. If these responsibilities were federalized, however, then resource limitations would prevent immediate attention to all needs. Efforts to prioritize coastal ports in terms of their strategic or economic importance would be politically perilous. Given the diminishing availability of federal funds for public works projects and the movement to shift responsibilities to the states, it has been convenient to

---

[7] The U.S. Congress recognized as early as 1971 that the acquisition of suitable disposal areas for the significant quantities of materials dredged in the course of the nation's navigation projects was a significant problem. Efforts were made to address the problem in several areas of the country. A program was established to construct land placement areas for Great Lakes projects. The Port of Baltimore embarked on a plan to build an upland disposal area for sediments from Baltimore Harbor channels. And the USACE began developing a comprehensive plan for extending the useful life of the Craney Island disposal area established in 1954 to service Norfolk Harbor channels.

avoid addressing the growing need for federal involvement in locating and funding containment facilities for dredged material.

The committee can offer no easy answers to this problem. A partial solution may be provided, however, by the trend toward the development of dredged material management plans (DMMPs). The USACE policy on DMMPs was established recently in Engineer Circular 1165-2-200. USACE regulations require that all navigation, federal harbor, and inland waterways projects have DMMPs that satisfy long-term needs for the management of dredged material. The objective is to establish project-specific plans (longer than 10 years) for the placement or management of dredged material consistent with applicable laws and policies. The regulations make provisions for the local requirements of ports and harbors. DMMPs are to be carried out in cooperation with project sponsors, local governments, port authorities, and other project users and beneficiaries. The regulations encourage beneficial uses of dredged material and outline the procedural requirements for managing certain dredged materials authorized by recent versions of the WRDA. The regulations do not address the management of contaminated sediments in detail, but DMMPs can include consideration of technologies for the treatment and management of contaminated sediments. The National Dredging Team soon will issue parallel guidelines for the development of long-term DMMPs that will complement the USACE regulations and directly involve the federal and state agencies serving on the regional dredging teams.

STAKEHOLDER INTERESTS

Contaminated sediments are not managed in a political or social vacuum. Most contaminated marine sediment sites are located in highly populated areas. The location of these sites virtually ensures that project proponents must contend not only with complicated ecological situations and difficult technical problems but also with complex political circumstances involving multiple resource users and interest groups. The selection of ex situ disposal or containment sites usually affects stakeholders. As a result, successful management of any contaminated sediments problem must respond to all dimensions of the problem: ecological, technical, and political. This chapter examines the strategic and diplomatic skills required of project proponents and the tools available to them. The committee gained considerable insight into these issues through the case histories (see Appendix C).

Stakeholder Groups

The stakeholder groups that need to be considered for possible inclusion in a the decision-making process are diverse, although they often have common interests, such as economic or environmental concerns. Stakeholder groups include local communities, which could be exposed to contamination either in the sediments or at a placement site or which could depend on a contaminating industry for employment. The stakeholders might also include communities that are dependent on fisheries where contaminated sediment is found in a critical habitat or along a migratory pathway. Local industries and ports also have an interest in sediment management because the regional economy may depend on a water transportation system for shipping manufactured goods; an attractive environment; or abundant stocks of healthy, edible fish or shellfish to support commercial and recreational fisheries. Riparian landowners have a stake in nearby dredging and disposal activities.

An important general category of stakeholders encompasses environmental and other public interest groups, whose concerns can vary widely. One group may focus narrowly on a local issue, another might take a regional perspective on growth and development, and a third might have a global environmental agenda and be targeting local entities, such as companies that make pesticides or refine hydrocarbons into fuels. Some groups have specific interests in one ecosystem component, such as endangered species, migratory birds, or non-native organisms.

As more citizens become aware of and are educated in marine and coastal issues and seek to participate in decisions to allocate resources, the list of stakeholders grows. Although these diverse interest groups initially may hold widely varying positions on contaminated sediment issues, including being wholly misinformed about the range of management issues, they must all be considered to ensure public acceptance, expedite action, and maximize prospects for long-term success.

Phases of Involvement

Chances for successful site management are enhanced if stakeholder involvement begins early and continues throughout the decision-making process. The design of the decision-making process and the initial efforts to develop that design are key avenues for injecting stakeholder values—whether based on economic self-interest or a more elusive "public" or "community" interest—into the development of a mutually acceptable solution. The Port of Tacoma case history demonstrates the benefits of bringing all stakeholders into the process from the start and forging a collective solution. As stakeholders participate in management processes and are exposed to the full spectrum of facts and viewpoints, their values evolve, not only throughout the decision-making process but also, perhaps more critically, during project implementation.

It is especially important that stakeholders be involved in the identification and selection of management strategies and the concomitant weighing of risks, costs, and benefits. Major disagreements concerning the management approach need to be resolved at this stage to ensure that the chosen solution can be implemented without lawsuits or other delays. Dispute resolution is discussed in the following section.

Consensus Building

Early stakeholder involvement permits the various parties to feel included in the process, acquire mutual trust, and gain an in-depth understanding of the problem. These elements foster the development of a consensus, which is critical to the timely implementation of solutions. In the Port of Tacoma case, agreement between the EPA and the port helped satisfy regulatory, environmental, and local economic concerns and led to a successful outcome. Conversely, the absence of early cooperation among stakeholders was partly responsible for delays in the Boston Harbor and Hart-Miller Islands case histories. In the latter case, an attempt to rush the project through without identifying and resolving the controversy led to a legal challenge and inordinate delays. The case histories also indicate that successful consensus building depends on the emergence of a strong project proponent who is willing to assume responsibility for advocating key objectives, integrating the various processes, and resolving whatever conflicts arise.

Early involvement of stakeholders cannot guarantee success. Increasingly, contaminated sediments are being managed in complex, changing social and political settings marked by the emergence of nontraditional stakeholders. Conflicts are virtually inevitable in this context regardless of the quality of the decision and, when conflicts arise, they must be addressed directly through an appropriate conflict or dispute resolution approach. Professionals in resulting disputes helped stakeholders reach agreement in the Boston Harbor case.

A variety of decision-making approaches has emerged in recent years that can be used to help resolve disputes or conflicts. The simplest method is to bring stakeholders together for a frank, constructive discussion. Other approaches include mediation, negotiated rule making, and collaborative problem solving. Another approach that may help indirectly is decision analysis (a concept introduced in Chapter 2 and described in detail in Appendix E), which, when used appropriately to estimate the outcomes of particular management strategies, can provide insights that could help foster consensus. Each of these approaches has a place in the arsenal of techniques for improving the prospects for a politically acceptable, implementable decision. A detailed analysis of these techniques is beyond the scope of this report, but the key aspects are summarized below.

Whether carried out in the context of mediation, arbitration, or collaborative problem solving, fostering a consensus on a management process among all interested or potential stakeholders involves more than simply going through the mechanics of communicating with all parties. The case histories underline the paramount importance of developing positive working relationships that improve the chances of accommodating or resolving conflicts. The process thus requires not only knowledge and mastery of methodologies for building consensus but also interpersonal skills in establishing positive human relationships.

The literature on conflict resolution stresses that the handling of threshold questions is as central to success as the substantive outputs from the process itself. Threshold questions include who should be at the table, who should represent whom, how the interests of important stakeholders who do not come forward should be determined, how a common constructive definition of the problem(s) is to be developed, and how a mutually acceptable decision-making process is to be chosen. There is a significant body of literature on dispute or conflict resolution. Carpenter and Kennedy (1988) provide lay readers with an extended discussion of the mechanics of a powerful dispute resolution program, many examples of public dispute resolution, and a detailed bibliography (see also Singer, 1990).

Conflict resolution techniques have been used in some contaminated sediment cases, but the frictions that continue to plague and delay some projects indicate that they could be used much more. Federal agencies are authorized and encouraged to engage in alternative dispute resolution techniques by the Administrative Dispute Resolution Act of 1990 (P.L. 101-552). The USACE has developed guidelines for using these techniques to resolve contract disputes but has not formalized their use in contaminated sediments situations. The EPA frequently uses formal dispute resolution techniques, both in developing regulations and in dealing with specific Superfund projects.

While seeking to resolve disputes, it is important to remember that consensus building takes time, and time is always limited. Moreover, projects cannot be designed based on lowest-common-denominator choices. Often stakeholders are unwilling or unable to move beyond a certain point—because of deeply held principles, rigid legal restrictions, or budgetary limitations. In these cases, decisions must be made, even if disagreements remain.

Risk Communication

The management of contaminated sediments requires the active participation of diverse stakeholders from the onset of the decision-making process, even though conflicts are virtually unavoidable and, therefore, must be addressed directly. To ensure that a specific management alternative satisfies the concerns of all parties and can be implemented without unnecessary confrontations among stakeholders, they must be convinced to "buy in" to the credibility of the process in place, particularly with respect to the following issues: that corrective action will attain at least minimum acceptable risk levels; that proven methods for managing residual risks will be used; and that a process is in place for balancing risks, costs, and benefits in strategy selection and implementation.

Stakeholder buy-in can be fostered through risk communication, which is an integral part of risk management. Risk communication is defined as the exchange of information and opinions about risks among concerned individuals, groups, and institutions (NRC, 1989). As discussed in Chapter 2, interaction between the risk assessor and the risk manager is one aspect of risk communication; stakeholders become involved at the next level of the process, when the risk manager communicates with the public (USACE, 1991). Through effective risk communication, stakeholders' understanding of project-specific issues can be improved substantially and may result in a consensus on the choice of a specific course of action.

Risk communication, coupled with an orderly planning framework (as described in Chapter 2), can enhance and expedite the decision-making process. Through risk communication, factual and subjective information can be organized, evaluated, and communicated to stakeholders so that they become vested in the management process.

Beneficial Uses

Dredged sediments traditionally have been viewed as waste. However, dredged material is often used for beneficial purposes—fill for urban development (such as the construction of National Airport in Washington, D.C.), for beach nourishment, for the creation of wetlands and wildlife habitat, for improving farmland, as fill for general construction, and for establishing coastal islands where many species of seabirds nest (NRC, 1992). Dredged material can also be used as cover for sanitary landfills, caps for more-contaminated materials, and bases for underwater berms and breakwaters. These and other projects have helped meet the growing demand for placement sites for dredged material and have yielded economic and environmental benefits as well. The statutory underpinning for the beneficial use of dredged material is provided by WRDA 1992 (P.L. 102-580), which contains provisions for using dredged material for such things as the protection, restoration, and creation of aquatic habitat.

Most of the sediments put into beneficial use have been "clean," but contaminated dredged materials can also be reused safely. The most straightforward method is to isolate the contaminated materials from the surrounding environment. For example, contaminated dredged material can be placed in the interior of a diked containment facility, which can then be capped with clean materials. This approach was used in the Hart-Miller Islands project, where the containment facility was used as the foundation for a recreation area. Similarly, in the Port of Tacoma project, two-thirds of a secondary channel was filled with dredged material (much of it contaminated) to create 24 acres of land for the expansion of a marine container terminal and for habitat restoration.

Research has shown that some contaminated materials can be reused safely without being completely isolated from the surrounding environment, as long as the site is managed properly. Heavy-metal-contaminated dredged material from Black Rock Harbor, Connecticut, was used to create a wetland, which was gradually covered with dense vegetation. Tests on plant tissue and local snails revealed levels of heavy-metal concentrations similar to the level in the surrounding region (Francingues et al., 1996). A 46-acre CDF at Times Beach, New York, partially filled with contaminated dredged material in the early 1970s, was rapidly colonized by plants, animals, and birds and was designated as a nature preserve (Stafford et al., 1991). Numerous studies have been conducted on the fate of the heavy metals and organic contaminants at the site, yielding data that can be used to improve CDF management in general (Stafford et al., 1991, and references therein). In another example, contaminated dredged material from the Calumet River was used to restore an acid coal mine tailing area at Ottawa, Illinois, in 1978. Three feet of dredged material was placed on top of the mine tailings (which had a pH of 3.0) to correct the acid runoff problem and allow vegetation to stabilize the restored site. Agricultural crops were grown on the dredged material that were of equal quality and yield with crops on surrounding farms.

Beneficial uses of contaminated dredged material are, however, comparatively rare. The reasons for this are not entirely clear but probably include public resistance (or the fear of public resistance) to the reuse of contaminated materials; the USACE policy of pursuing the lowest-cost environmentally acceptable dredged material placement alternative (33 CFR §335.7); and the recognition that the benefits of reuse often accrue to third parties, whereas the added costs must generally be borne by the project sponsor. The cost issue has been particularly contentious. Ports and the USACE have favored assigning the incremental costs of beneficial uses to the beneficiary, whereas states have tried to compel the USACE to bear the costs by requiring beneficial uses under the authority of the CZMA (Coastal Zone Management Act) (Maritime Administration, 1994). Conflicts like these have delayed projects and driven up costs (Maritime Administration, 1994).

Nevertheless, there are sound reasons for encouraging the beneficial use of contaminated sediments, particularly at this time. First, beneficial use can improve the cost-benefit outlook for managing dredged material (see Chapter 2 and Appendix D) because it not only eliminates the need for costly conventional placement (typically far more expensive for contaminated than for clean sediments), but it also can provide economic benefits. Second, if properly handled, beneficial uses can foster public and political support for otherwise objectionable plans for the placement of dredged material, as demonstrated by the Port of Tacoma project, which is generally viewed as having been beneficial both to the port and to the environment. Third, new EPA regulations (40 CFR §503 [1994]) that promote the reuse of sewage sludge by significantly increasing permissible levels of most contaminants are being used by the USACE to evaluate data on, and the reuse of, dredged material (C.R. Lee, USACE, personal communication to Marine Board staff, December 15, 1995).

Although there are no guidelines for the reuse of contaminated sediments, limited research, prompted by the shortage of storage space for dredged material and the new EPA regulations, is under way on the safe, beneficial use of contaminated material. Over the past few years, the USACE has been working on a variety of projects with industry and universities using dredged material with various levels of contamination. One project, for example, is "manufactured" soil. Another project has focused on whether silt can be replaced with lightly contaminated dredged material from a CDF (in Toledo, Ohio) to compensate for a shortage of silt, which is normally combined with peat and organic yard waste to make topsoil (C.R. Lee, USACE, personal communication to Marine Board staff, December 15, 1995). The Toledo material already has been used to make landfill cover (in a ratio of 90 percent sediments, 8 percent sewage sludge, and 2 percent lime to fix metals). Another set of projects with the USACE New York District is evaluating the use of more highly contaminated sediments as well as post-treatment residues for use as soil, road aggregates, and building blocks (C.R. Lee, USACE, personal communication to Marine Board staff, December 15, 1995). Although this research is promising, funding is limited. A new USACE five-year Dredging Operations Environmental Research program, focusing specifically on the management and reuse of contaminated materials, has been proposed but has not been funded.[8]

Several steps could be taken to promote the beneficial use of contaminated dredged material. First, the economic and social acceptability of beneficial uses are likely to be greatest if these alternatives are considered as part of a generally accepted package of solutions, rather than as one-shot experiments. This is one of the potential advantages of long-term management planning—including long-term permitting—for ports (Interagency Working Group on the Dredging Process, 1994), an approach the USACE is adopting. At the same time, economic incentives could be created if federal policies were modified (or existing policies implemented) to encourage the beneficial use of contaminated dredged material, even if beneficial uses are more expensive than the lowest-cost, environmentally acceptable placement. Either the requirement to use the lowest-cost alternative could be waived for alternatives that include beneficial uses (difficult during a period of fiscal austerity) or the USACE could modify the way it computes initial costs by treating the economic benefits of beneficial uses (including elimination of the need for other placement sites) as cost offsets. Over the long term, the advantages of such approaches could outweigh the additional costs associated with the implementation of beneficial uses.[9]

Second, research on the beneficial uses of contaminated dredged material seems particularly timely, given the growing need for cost-effective placement

---

[8] The New York District work includes cooperative research and development agreements (CRADAs) with several private companies but has a limited budget, and the Toledo project has an even smaller budget. Much of the funding is spent on chemical analyses. Under CRADAs, companies provide materials (such as organic waste materials), the local USACE district office supplies the dredged material and the funding, and the USACE Waterways Experiment Station in Vicksburg, Mississippi, conducts the tests and evaluations of various mixtures of materials. The rules for CRADAs allow companies to provide funding (63 USC 15 §3710a[d1] and U.S. Army Regulation 70-57), but no private dollars have been provided to date for research on the reuse of contaminated dredged material.

alternatives and the promising results of research. Although a detailed analysis of funding is beyond the scope of the present report, there would seem to be a persuasive argument at least to continue, if not expand, current research efforts.

Off-Site Mitigation

Political tensions can be compounded by the reality that no contaminated sediment management strategy can be 100 percent successful. It is not possible to remove all contamination from a site, to dredge a channel without causing peripheral impacts, or to guarantee that a suitable placement area can be found without causing dislocations. Lost value or unavoidable impacts may be offset, however, through off-site mitigation, which involves providing alternative resources to injured parties. In the Waukegan Harbor case history, for example, the containment area selected for the contaminated sediments was occupied by a recreational marina. Public objections to the loss of recreational waterfront were relieved when the marina was relocated to a site donated by the company responsible for the contamination. In this way, the public was compensated for dislocations caused by the sediment placement strategy.

The mitigation approach could be used just as easily to offset incomplete remediation. Some areas beset with chronic, widespread contamination are unsalvageable; natural functions may be "restored" (i.e., to a degree deemed adequate) over a very long period of time—or not at all. In such cases, the environment could be "made whole" through the establishment and preservation of an area of comparable or greater value at another location. This is a long and complex process, however. Because the development of a properly functioning ecological habitat is an imperfect science, the process is often carried out by trial and error, and it may take decades. More than one acre of replacement habitat (e.g., in

---

[9] Several provisions of WRDA 1986, 1990, and 1992 take steps in this direction, although they do not address the issue of contaminated sediments specifically, and they are seldom used. For example, Section 145 of WRDA 1976, as amended by Section 207 of WRDA 1992, authorizes the USACE to enter into agreements with states or localities to use dredged material for beach nourishment. This section does not address the funding issue, and it applies only to beach-quality sand. Section 907 of WRDA 1986 deems "environmental quality enhancement" benefits to be at least equal to costs in the cost-benefit ratio comparison. This provision applies only to new-work dredging projects when cost-benefit ratios are used. Section 906 of WRDA 1986 provides authority to mitigate fish and wildlife losses resulting from a water resource project at 100 percent federal "first costs," when the benefits are national, or at 75 percent, when they are more localized. This provision is limited to mitigating damages caused by the project. And Section 204 of WRDA 1992 authorizes the USACE to undertake projects (usually at 75 percent federal funding) for the protection, restoration, or creation of aquatic and ecologically related habitats, when justified by the "monetary and non-monetary" benefits. This section is limited to discrete projects with specified beneficial use objectives; thus, navigation dredging projects would not normally qualify.

wetlands for example) is typically required to offset an acre's worth of environmental damage (NRC, 1994). Newly established habitats must be monitored for a prolonged period to ensure that the desired ecological functions have been restored. Fall-back measures must be developed in case initial efforts are unsuccessful, and appropriate legal safeguards must be established to ensure that mitigation measures cannot be undone and that newly created areas are not contaminated or disturbed in the future.

SUMMARY

A number of important findings emerged from the committee's analysis of the regulatory framework for the management of contaminated sediments and the many issues related to stakeholder interests. It is clear that certain aspects of current laws and regulations may impede the cost-effective management of contaminated sediments. The committee identified three areas in which improvements are both warranted and possible.

First, laws and regulations tend to emphasize mechanics rather than balancing risks, costs, and benefits. None of the three laws (MPRSA, CWA, and Superfund) governing the evaluation of remedial alternatives explicitly considers either the risks posed by contaminated marine sediments or the costs and benefits (i.e., economic risks and technical viability) of the possible solutions. Similarly, permitting processes for sediment placement focus on the location of the placement site and the reason for the dredging rather than on the risk posed by the contamination. In the committee's view, more consideration needs to be given in the regulatory process to risk.

Second, timely decision making and the implementation of cost-effective solutions may be impeded by too much reliance on procedures without regard for the intent of statutes. An objectives-based interpretation of regulations focusing on the underlying intent of the statute(s) may foster the implementation of the best management practices and creative solutions to difficult problems.

Third, regulations governing cost allocation and cost-benefit analysis for dredging and placement projects may foster unsound allocations of scarce resources. The federal government pays for most new-work dredging and all maintenance dredging, but the costs of sediment disposal are borne by the local sponsor, a requirement that creates a strong preference for open-water disposal and a disincentive to beneficial use. Furthermore, although costs and benefits must be weighed carefully for new-work dredging, similar cost-benefit analyses are not required for either maintenance dredging or dredged material placement.

There is also room for improvement in how project proponents interact with stakeholders. Failure to identify all important stakeholders early in the decision-making process and to build consensus among stakeholders can cause significant delays and even block the implementation of solutions. The development of consensus can be fostered by various consensus-building tools, including mediation, negotiated rule making, collaborative problem solving, and effective risk communication.

The beneficial reuse of contaminated sediments is attractive because it provides alternatives to increasingly scarce disposal sites and makes management plans more attractive, or at least palatable, to stakeholders. Some contaminated sites have been successfully transformed into wetlands, and productive research is under way on the safe use of contaminated sediments for various purposes. However, funding for this type of research is limited. Another barrier to beneficial reuse is the USACE emphasis on the selection of lowest-cost solutions with little regard to the monetary and non-monetary advantages of beneficial reuse.

REFERENCES

Carpenter, S.L., and W.J.D. Kennedy. 1988. Managing Public Disputes: A Practical Guide to Handling Conflicts and Reaching Agreements. San Francisco, California: Jossey-Bass.

Environmental Protection Agency (EPA). 1994. EPA's Contaminated Sediments Management Strategy. Draft. EPA 823-R-94-001. Washington, D.C.: EPA Office of Water.

EPA and USACE. 1991. Evaluation of Dredged Material Proposed for Ocean Disposal. Testing Manual. EPA-503/8/91/001. Washington, D.C.: EPA Office of Water and USACE.

EPA and USACE. 1994. Evaluation of Dredged Material Proposed for Discharge in Waters of the U.S. Testing Manual. Draft. EPA 823-B-94-002. Washington, D.C.: EPA and USACE.

Francingues, N.R., Jr., A. Brambati, H. De Vlieger, V. Haviar, Y. Hosokawa, H. Koethe, H.P. Laboyrie, E. Paipai, E. Van den Eede, C. Wardlaw, and W. Willemsen. 1996. Handling and Treatment of Contaminated Dredged Material from Ports and Inland Waterways-CDM, vol. 1, PTC 1, Report of Working Group No. 17, Supplement to Bulletin No. 89. Brussels, Belgium: PIANC General Secretariat.

Interagency Working Group on the Dredging Process. 1994. The Dredging Process in the United States: An Action Plan for Improvement. Report to the Secretary of Transportation. Washington, D.C.: Maritime Administration.

Maritime Administration. 1994. A Report to Congress on the Status of the Public Ports of the United States, 1992–1993. MARAD Office of Ports and Domestic Shipping. Washington, D.C.: U.S. Department of Transportation.

National Research Council (NRC). 1989. Improving Risk Communication. Washington, D.C.: National Academy Press.

NRC. 1992. Restoration of Aquatic Ecosystems: Science, Technology, and Public Policy. Washington, D.C.: National Academy Press.

NRC. 1994. Restoring and Protecting Marine Habitat: The Role of Engineering and Technology. Washington, D.C.: National Academy Press.

Singer, L.R. 1990. Settling Disputes. Boulder, Colorado: Westview Press.

Stafford, E.A., J.W. Simmers, R.G. Rhett, and C.P. Brown. 1991. Interim Report: Collation and Interpretation of Data for Times Beach Confined Disposal Facility, Buffalo, New York. Long-Term Effects of Dredging Operations Program. Miscellaneous Paper D-91-17. Vicksburg, Mississippi: U.S. Army Engineer Waterways Experiment Station.

U.S. Army Corps of Engineers (USACE). 1991. Risk Assessment: An Overview of the Process. Environmental Effects of Dredging. Technical Notes EEDP-06-15. Vicksburg, Mississippi: U.S. Army Engineer Waterways Experiment Station.

4

Site-Specific Considerations

An initial characterization of the source(s), type, and extent of site contamination and bioavailability needs to be conducted coincident with, or even before, an evaluation of regulatory realities is made and identification of the stakeholders and their particular interests is established. The range of factors governing transport and contaminant concentrations in marine systems requires that assessment procedures and methods be site specific. A reasonable understanding of site dynamics is also necessary to evaluate the proposed methods of characterization and methods of site assessment in terms of cost effectiveness and scope.

This chapter deals with contaminant sources, transport processes, and methods of site characterization. Site-specific considerations are important, in the committee's judgment, because inadequate source control and site assessment can undermine the best management practices. This chapter outlines how appropriate attention to these issues can help control costs and enhance the effectiveness of sediment management. However, the discussion is not intended to provide comprehensive, step-by-step guidance or to evaluate all methods that may be applicable. The emphasis is on the need for a systematic approach that couples site-specific information with remedial efforts. For each project, the time and resources required for source control and site assessment need to be weighed against the projected benefits of these activities, the availability of quantitative data, and the need to proceed with site management.

SOURCES OF CONTAMINATION

Source control refers to measures undertaken to identify and curtail continuing sources of contamination. Source control is advisable in all situations. It may be impractical in navigation projects but needs to be an integral component of environmental remediation projects, except in the most unusual circumstances. Failure to control the source of contamination leads to the recontamination of newly exposed sediments, in which case remediation efforts have to be considered unsuccessful.

Source control is not, however, easy or inexpensive. In some cases, contaminant source(s) cannot be identified. Even if they can be pinpointed, some types of contamination, such as atmospheric fallout, are difficult or impossible to control. Another difficulty is the question of who is responsible for source control. From the standpoint of both economics and fairness, the costs of prevention and control ought to be borne by the polluter(s) and internalized into their production costs. Indeed, U.S. environmental law is generally based on the principle that the polluter pays. But those responsible for sediment contamination are not always (and sometimes cannot be) held to that standard. Thus, under current regulations, the burden for source control is not distributed equitably, which means that some sources of contamination are not controlled at all.

Source control is used more often in environmental remediation projects, which are usually funded by the government (i.e., taxpayers), than in navigation dredging projects, which are partly financed by commercial navigation users through various tax assessments established by the WRDA of 1986. In Superfund site cleanups, a legal mechanism may be available to force upstream sources of contamination to bear an appropriate share of remediation costs and even to require the abatement of ongoing releases. However, in navigation dredging projects, the local port authority or other dredging proponent usually has little leverage over upstream polluters and, in the case of atmospheric deposition, virtually none over polluters outside the watershed. Thus, contamination may persist, leading to a continuing need to dredge and redredge contaminated sediments, which is costly and politically unacceptable.

Source control could be encouraged in navigation dredging projects through regulation, as long as the question of who pays is resolved in a manner that is acceptable to all parties. A port cannot be expected to finance source control as well as sediment remediation (allocation of remediation costs is discussed in Chapter 3) when it is not responsible for the initial contamination. The primary focus needs to be on the development and implementation of state and federal pollution prevention programs aimed at reducing or eliminating the sources of sediment contamination.

Regulators have long recognized that the identification of upstream sources of contamination is essential for the progressive improvement of water quality. CWA §303 emphasizes control of point-source discharges using technology-based measures but allows especially stringent discharge limits to be imposed by states based on the water quality in a given area. The logic of this approach applies equally to contaminated sediments.

Section 303 could be amended to require EPA and delegated states to consider the impact on the quality of sediment downstream in setting total maximum daily loads (TMDLs) for waterway segments and developing load allocations for contaminant sources.[1] In addition, congressional initiatives (i.e., CWA reauthorization legislation) could require watershed-specific inventories (including the identification of contaminant sources) of upstream contaminant contributions to sediment contamination downstream in port areas. In situations where watershed planning has failed and identifiable upstream sources have contributed disproportionately to sediment contamination downstream, the EPA could be authorized to recover an appropriate share of cleanup or disposal costs from the responsible parties.

Part of the EPA's draft document on a strategy for managing contaminated sediments (EPA, 1994) outlines the agency's use of sediment quality criteria (SQC). One chapter describes how the water program (Office of Water) will permit municipalities and industrial facilities to meet SQC. The EPA has also initiated an inventory of sites and sources of sediment contamination using information from national databases.[2] These initiatives will be very useful, if not critical, to the understanding and control of sources of contaminated marine sediments.

CONTAMINANT TRANSPORT AND AVAILABILITY

Decision makers must understand the factors affecting contaminant transport and availability to develop a site characterization plan and, eventually, to evaluate site management alternatives. Understanding these factors can help minimize project costs, foster the development of efficient and effective sampling plans, and assist in the selection of optimum remedial schemes. This section outlines the primary factors.

The distribution of contaminants in the coastal marine environment is determined by complex interactions among meteorological, hydrodynamic, biological, geological, and geochemical factors. Interactions within and among these factors result in a transport system with wide variations, both spatial and temporal. This variability complicates site assessment surveys and requires that care be taken to specify the frequency and location of field samples.

Usually the time scales range from hours to months and are reasonably

---

[1] This approach, although it might be difficult to implement, could be designed to address sources of sediment contamination. Although resolution of source control problems is outside the scope of this report, these issues warrant further attention.

[2] See the National Sediment Contamination Point Source Inventory: Analysis of Facilities Release Data for 1994 and National Sediment Quality Survey: A Report to Congress on the Extent and Severity of Sediment Contamination in Surface Water. Both documents are in development as of this writing.

regular. The patterns are sometimes disturbed, however, by high-energy storms, which can displace large amounts of sediments and significantly alter the distribution and availability of contaminants. Thus, comprehensive site assessments need to include consideration of the effects of both long-term, periodic variations and infrequent, but often high-energy, aperiodic events.

Beyond the issue of spatial and temporal variability, assessment of coastal marine sites can be further complicated by the inherently nonlinear behavior of the transport system affecting the distribution and availability of contaminants. System response seldom displays simple functional dependence on force magnitude. Thus, evaluations typically need to consider other factors, such as the history of disturbances or antecedent conditions. For example, the effects on water-column mixing, of winds of identical velocity and duration vary greatly depending on the direction and magnitude of tidal conditions. These and other nonlinear tendencies are particularly pronounced in the processes that govern the transport of fine-grained, cohesive sediments.

Fine-grained silts and clays (inorganic particles less than 60 micrometers in diameter), because of their relatively large surface-area-to-volume ratio and electrochemical character, are the favored adsorption sites for most contaminants found in coastal areas (Gibbs, 1973; Moore et al., 1989). These sediments enter the system from a variety of local, upstream, and offshore sources and can be transported initially as discrete particulates suspended in the water column. Larger, sand-sized particles are moved closer to sources by sedimentation, whereas fine-grained particles are readily dispersed.

With time, individual particles come together to form larger-diameter aggregates as a result of either physicochemical coagulation or biologically mediated agglomeration. In the water column, the sizes of these aggregates and their associated settling velocities are controlled by the balance between collision and breakup forces induced by flow-associated shear. This force balance continuously changes as the particles migrate through differing flow regimes caused by horizontal advection and turbulent mixing.

The process continues as long as flow energy and the associated boundary shear stresses are high. As energies decrease (as in many estuaries and dredged channels), aggregates settle to the sediment-water interface, forming a loosely consolidated, high-water-content surficial deposit (NRC, 1987), often referred to as a "fluff layer." This layer is highly porous with minimal shear strength, so subsequent tidal cycling typically results in the resuspension of some or all of the deposit, favoring low rates of net deposition. Cohesive sediments tend to consolidate slowly because of the weights imposed by the cyclic loading of surficial materials and because of a response to the increasing burden imposed by persistent net deposition acting in combination with the varying surficial load. This process favors the development of a column of sediment in which physical strength and associated erodibility vary significantly with depth.

In addition to physical loadings, the vertical structure of the sediment column in a fine-grained deposit is affected by a variety of chemical and biological factors. The sediment-water interface represents a relatively distinct chemical boundary separating the generally oxygenated water column from an anoxic sediment column. This transition, typically occurring within a few centimeters of the interface, favors reducing conditions within the body of the sediment column and the dominance of facultative and anaerobic bacteria. Changes in pH (an indicator of acidity) and Eh (a measure of oxidizing potential) associated with this transition can directly affect sediment contaminant availability, altering the degradation of organic matter and providing a sink for selected trace metals. The latter process can be particularly pronounced in sulfate-rich seawater, resulting in the precipitation of trace metals by sulfides in the anaerobic pore waters and the subsequent down-gradient diffusion from surficial, aerobic sediments to the deeper anoxic pore waters.

The rates of degradation of organic matter are also affected by the shift from oxidizing to reducing conditions within the upper levels of the sediment column (the redox gradient), with more effective microbial degradation of bioavailable compounds of concern (e.g., polyaromatic hydrocarbons) possible within the oxic region. These processes slow significantly within the deeper anoxic regions of the sediment column, often resulting in contaminant half-lives on the order of years.

The tendency of fine-grained materials to assimilate and concentrate nutrients and organic substances attracts a diversity of macrobiota, particularly within the upper 20 to 40 centimeters (cm) of the sediment column. The activities of these deposit and filter-feeding organisms significantly modify sediment fabric by burrowing and altering surface roughness, internal porosity, and physical strength. These modifications, known as bioturbation, can be expected to alter contaminant transport pathways and the overall erodibility of the sediment deposit.

The combination of physical transport, chemical interactions, and biological processing results in sediment deposits typically characterized by horizontal gradients that are weaker than vertical gradients. The depositional sequence described above favors the formation of a mobile, near-surface layer of material overlying a reasonably well-consolidated and virtually immobile interior (Hayter, 1989; Ross and Mehta, 1989; Bohlen, 1993). The mobile layer, which is subject to diffusive or advective processes, is generally confined to the immediate sediment-water interface and is seldom more than 2 to 4 cm thick. Boundary shear stresses produced by the prevailing flows are sufficient to displace only the upper portions of this region, including the fluff layer and a thin underlayer no more than 1 to 2 millimeters (mm) thick. Displacement of the entire mobile layer requires boundary shear stresses that occur only during major storms. The deeper interior region, below the mobile layer, is even more resistant to transport and can be considered immobile in the absence of loadings extreme enough to produce mass failure of the entire deposit. This vertical gradient in erodibility has profound implications for evaluating sediment-associated contaminant availability, particularly in projects where natural remediation can be considered.

The variety of factors affecting sediment erodibility makes it difficult to predict the response of a given deposit to a specified range of forces. Deposition rates, chemical environment, and biological activity can vary significantly, both spatially and temporally. This complexity directly affects the fabric of the sediment column and typically precludes the development of a generally applicable transport algorithm. As a result, erosion rate models require site-specific data. The application of site-specific formulas can be complicated further by the sensitivity of a given region to disturbances. Typically, the first storm of the season, acting on a sediment surface formed during an extended period of low transport energy, displaces a significantly larger mass of sediment than subsequent events, despite similarities in peak energies. These differences in response are often difficult to specify quantitatively, complicating the development of predictive numerical models for site assessment or management.

SITE ASSESSMENT: APPROACH, METHODS, AND PROCEDURES

The variety of factors affecting the distributions and availability of sediment-associated contaminants require the site manager to use a well-structured, tiered approach to site assessment (see Figure 4-1). (This approach expands on the one outlined in Chapter 2, Figure 2-1.) The remainder of this chapter outlines the elements of this approach. The committee views this kind of approach as having the best potential for achieving overall cost effectiveness and for clearly focusing on survey and remediation efforts. Focus means having a clear definition of project objectives, the satisfaction of which is the sole purpose for acquiring survey data, characterizing contaminant distributions and availability, and designing and selecting remedial schemes. None of these activities is an end in itself; each is justified only to the extent that it contributes to the fulfillment of project objectives (see Box 4-1).

Use of Historical Data

To ensure the cost-effective management of contaminated sediments, site characterization needs to begin with a review of the past and present uses (residential, commercial, and industrial) of waterways and adjoining lands. An understanding of past uses can place some bounds on the range of contaminants stored within the sediment column and highlight important geographical or archeological features of the site. The knowledge of present contaminant discharges and local transport dynamics can provide an immediate indication of the long-term effectiveness of a contaminant removal strategy and the overall advisability of proposed uses of the site. Source control is an important element in the management of contaminated sediments. Acquisition of historical data requires a careful

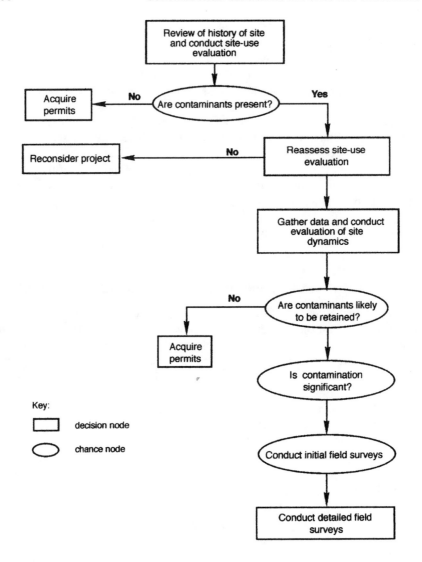

FIGURE 4-1 Conceptual site assessment protocol.

search of a variety of repositories, including state and federal permit files and water quality data; municipal planning, zoning, and land-use records; and assessors' files of deeds and titles dating back to the preindustrial period. Although data gathering requires resources, failure to identify the historical features of a site can also result in wasted time and money. This lesson was evident from the

> **BOX 4-1**
> **Basic Tenets of Site Assessment**
>
> After 20 years of cleanup experience, some basic tenets of site assessment have emerged. The committee developed the following list:
>
> - An understanding of site history, existing conditions, and dynamics is needed for the design and implementation of a successful management plan.
> - The process of site assessment is complex and expensive, but it is possible to obtain the information necessary for making informed decisions.
> - There is always some uncertainty associated with any decision; if one waits until all uncertainty has been eliminated, then no decision will ever be made.
> - Data gathering must focus on meeting specific needs; data gathering is not an end in itself.
> - Good site assessment results in minimum-cost projects that meet cleanup objectives.

Marathon Battery case history, in which remediation plans had to be redesigned to accommodate the late discovery of an old gun-testing platform (see Appendix C).

The area to be included in the historical survey depends on the transport dynamics and routes, both hydrologic and atmospheric, that affect the contaminants of concern. The project area may extend well beyond the immediate confines of the site, out to, and sometimes beyond, the boundaries of the watershed. An understanding of past operations affecting the area and the range of previous management concerns can define the character and loadings of a contaminant and can provide a qualitative indication, at least, of probable areal distributions, both horizontal and vertical. This information is essential to the design of subsequent field surveys and, if carefully gathered, can increase significantly the cost effectiveness of field surveys.

The experience of committee members suggests that the value of historical reviews for enhancing cost effectiveness is often overlooked. Historical data may be set aside on the assumption that past analyses do not meet current standards. Although quality assurance and quality control problems may limit the value of quantitative historical data, several recent studies suggest that they do not justify outright rejection. In the ongoing cleanup of Boston Harbor, for example, the U.S. Geological Survey (USGS) reviewed available historical data in an effort to increase the resolution of the sediment characterization study. Although the majority of these data would have failed current quality assurance/quality control criteria, computer-based batch-screening procedures keyed to internal consistency and wild point minimization allowed recovery of nearly 3,000 useful data points (F. Mannheim, USGS, personal communication to Marine Board staff, November 5, 1995). This figure represents a sixfold increase in the original data file. The resulting improvement in data density may significantly improve qualitative evaluations of the degree of contamination, complementing evaluations of toxics transport and flux and assessments of the natural recovery rate of the system.

Historical site data, combined with reviews of governing regulations and stakeholder interests and the definition of ongoing contaminant discharges, enhance the quality of the evaluation of the proposed uses of the site and the scope of the associated management efforts. Consideration of these issues complements the specification of remediation end-points and the criteria for field surveys. Honest, reasoned considerations of site use can provide early indications of project advisability, save significant time and money, and foster goodwill. For example, early reviews may indicate that expansion of port facilities is inadvisable because the area is characterized by high sedimentation rates that require frequent dredging and that efforts would better be directed or confined to another site that needs less maintenance.

Evaluation of Site Dynamics

After a consideration of site use, the next step is an initial evaluation of site dynamics. The purpose is to determine the extent to which contaminants entering the study area are retained and the probable location of major repositories or sinks. Because most contaminant transport is associated with the displacement of fine-grained sediments, these evaluations place particular emphasis on factors that affect sediment erosion, transport, and deposition. Depending on the location, it may be possible to define the majority of these factors using existing information. Data are needed concerning the topography of adjoining lands and ground cover characteristics; local tidal height and currents; stream flows; meteorology, particularly wind speed and direction and concurrent air temperatures; water depths; and surficial sediment characteristics. Data should extend over the range of seasonal conditions and include indications of system response to aperiodic storms.

The majority of these data can be obtained from federal agencies, including the National Weather Service for meteorological data, the National Ocean Survey for tidal and bathymetric observations, the U.S. Geological Survey (USGS) for stream flows and regional topography, and the Soil Conservation Service for ground cover characteristics. Other information, including data on surficial sediment, may be available from the USACE, the EPA, or local, state, or municipal regulatory groups or agencies, such as departments of transportation responsible for the construction and maintenance of road and railway bridges. The latter group is an often-neglected source of information. Foundation designs for roads and bridges often require deep, drill-hole data detailing soil characteristics. These data often provide a unique view of the vertical structure of the sediment column down to bedrock at several points across a waterway. Such perspectives are difficult and expensive to obtain but, when available, the may be of great value in investigations of sediment transport dynamics and associated contaminant availability.

Transport data (e.g., tides, winds, and stream flows), in combination with information detailing municipal and industrial outfall locations and numbers, permit an initial evaluation of the probability that contaminants and sediments entering the waterway may be retained, in whole or in part, within the project site. These evaluations place particular emphasis on surficial sediment characteristics and water depths. Contaminant retention within a basin where most sediments are sands or other coarse materials is likely to be less than within a basin dominated by fine-grained materials, all other factors being equal. Similarly, retention within shallow basins with smooth, regular features is generally less than in deeper systems, particularly those with abrupt discontinuities in water depth. The remaining factors detailing the regional flow regime permit evaluation of the sensitivity of the transport system to changes in meteorological and hydrological conditions, particularly aperiodic storms.

Initial site evaluation, in combination with data detailing sediment contamination, enables decisions to be made concerning the need for and form of supplemental field surveys. It is possible, although unlikely, that the initial evaluation of the transport system will indicate minimal contaminant retention, which would eliminate the need for additional surveys. More often, initial evaluations highlight deficiencies in the existing data and help define the most probable sites of contaminant retention within the project area. This is extremely valuable information and, if carefully developed, can reduce field survey costs significantly.

Field Surveys

Initial field surveys[3] are intended to address any obvious deficiencies in data indicated in the preliminary reviews and to provide quantitative data on the extent and character of contamination in the project area. If no site information is available, then the field work typically begins with a survey of water depths and evaluations of surficial sediment characteristics. Surveys focus on areas adjoining known or suspected contaminant outfalls. Depths typically are measured acoustically along surveyed transects.

Gross characterizations of sediment might be made with simple rod probes to "feel" the bottom, supplemented by occasional mechanical grabs to recover

---

[3] The discussion of field surveys relates to both navigation and environmental cleanup projects in which contaminated sediments are present.

masses of sediment for laboratory analysis. Alternatively, surficial sediment characteristics might be mapped acoustically and verified by direct mechanical sampling or visual surveys. The latter technique has the potential of providing high-resolution spatial coverage of both water depth and surface sediments, thereby significantly reducing survey time and costs.

The initial sampling locations are selected based on the definition of surface characteristics of the study area and adjoining outfall locations, as well as the purpose of the project (e.g., environmental cleanup, port maintenance, new construction). Field surveys need to focus on both depositional and sensitive areas. The intent is to characterize the degree and type of contamination in the area so as to provide a basis for evaluating the need for more detailed, high resolution, higher-cost surveys. Sites may be located within expected contaminant source or sink areas, in channels or slips to be dredged, and at one or more points upstream and downstream of the limits of the project area. In addition, areas where there are abrupt changes in sediment character, in shoreline use, or in the hydrodynamic regime typically warrant sampling. The initial sampling locations are generally selected through a collaborative effort by representatives of a variety of regulatory agencies and the project applicant or site manager.

At each designated location, core samples of the vertical sediment column can be obtained by a variety of mechanical methods (e.g., push corer, gravity corer, box corer, vibra corer). Typical sampling depths are on the order of 1 to 2 meters. In dredging projects, the vertical extent of the desired dredging establishes the required length of the sediment core. Core samples are retained in contaminant-free liners and returned to the laboratory for analysis. The number of samples extracted from an individual core is generally a function of the extent of stratification. If the sediment column is relatively homogeneous, then the entire core often is mixed to produce a composite, resulting in a single subsample for analysis. Significant stratification over the vertical tends to limit compositing to individual strata, and the number of samples generally is at least as high as the number of strata.

Analyses of samples typically include qualitative visual logging to detail stratification and grain size and a number of quantitative physical and chemical analyses. The character and extent of these bulk-sediment analyses depend on the project. In dredging projects, analytical protocols follow the guidelines specified in the so-called Green Book (EPA and USACE, 1991). Simple sediment quality surveys might be less comprehensive, with the analysis focusing on a particular contaminant or class of contaminants.

Bulk testing of composite sediments provides an indication of the "average" degree of contamination in the study area. The mixing in compositing is considered representative of the mixing that occurs during dredging. The procedure is not intended to yield high-resolution spatial data detailing contaminant concentrations throughout an area. Efforts to obtain such data would be initiated only if justified by the results of the initial field survey(s).

After the initial field survey(s) and the associated laboratory analyses, including evaluations of contaminant concentrations and biological availability, the character and extent of sediment contamination in the study area can be determined. If the results indicate contaminant concentrations and/or bioavailability above defined "action levels," or if there are special stakeholder concerns, then a more detailed survey of contaminant distributions may be required.

Detailed site surveys usually place particular emphasis on a single contaminant and require more dense physical sampling than the initial survey(s). Although physical coring is the most common and reliable method for detailed mapping of contaminant distributions, it is a slow and expensive process and, depending on the heterogeneity of the sediment column and the number of potential contaminant sources, it tends to provide limited spatial resolution. Contaminant concentrations are often interpolated horizontally, resulting in an overestimation of the mass or volume of sediment that needs to be removed. It is important, therefore, to develop and implement more cost effective site assessment technologies to replace physical coring.

Recent Survey Innovations

In recent years, several systems have been developed that appear to have the potential to supplement and, in some cases, replace physical coring. The most promising extends acoustic sub-bottom profiling techniques to permit high-resolution mapping of acoustic reflectivity. The resulting data can be related directly to a variety of geotechnical properties, including porosity, bulk density, and grain size. Samples from physical coring are still necessary as a baseline, but these new systems have the potential to reduce overall project costs and significantly increase the spatial resolution of field surveys. Increased resolution is needed if the full value of precision dredging technologies (described in Chapter 5) is to be realized.

Tests conducted by the USACE (1995) have shown that acoustic profiling techniques can provide accurate, high-resolution characterization of both surficial and sub-bottom sediments (McGee et al., 1995). At the current stage of development, acoustic profiling cannot identify chemical contaminants or measure their concentrations in sediment. However, acoustic profiling surveys can help define the thickness and distribution of disparate sediment types, as was demonstrated in the Trenton Channel of the Detroit River (Caulfield et al., 1995). In this demonstration, acoustic profiling produced the following results:

- Contaminated sediment surface reflection coefficients exhibited high spatial variability. This variability added to the natural heterogeneity of the fine-grained sediment deposits and increased the resolution required of the field surveys.

- Spectral changes in reflectivity occurred across the boundaries between contaminated and uncontaminated sediments.

In short, the results suggested that the acoustic signature of contaminated sediments differs substantially from that of uncontaminated sediments. Committee members had different views regarding the potential of acoustic profiling for differentiating between contaminated and uncontaminated sediments.

Initial research results suggest that the use of acoustic profiling in site assessment, combined with precision dredging, has the potential to reduce the costs of contaminated site remediation. In addition, acoustic profiling might be more effective and cost less than current techniques for evaluating and monitoring capped contaminated sediment sites. However, the ultimate utility of acoustic profiling remains to be demonstrated. Questions remain about how effectively acoustic techniques will be able to identify specific compounds and their concentrations in sediments containing a wide range of contaminants (N. Francingues, USACE, personal communication to Marine Board staff, December 15, 1995).

Complementing the development of remote, in situ sensing of physical sediment properties is a growing interest in the use of real-time or near-real-time chemical sensors for use in the field. These sensors can provide both point measurements and long-term, time-series observations. Although the majority of in situ sensors currently under development are intended for surveys of soils or groundwater (e.g., Lieberman et al., 1991; Apitz et al., 1993), many could be adapted for use in the marine environment. Sensors that measure pH, Eh, and pore pressure are already used routinely. Microelectrodes, providing millimeter-scale measurements of pH, oxygen, carbon dioxide, and ammonia, are also commercially available. Currently, these sensors are not capable of measuring contaminants of concern in sediments. Examples of near-real-time sensors include X-ray fluorescence for the detection of selected metals in sediments and mass-sensing piezoelectric transducers suitable for both solid and liquid phase determinations of a variety of contaminants, ranging from hydrocarbons to selected metals (Ward and Buttry, 1990).

Particularly promising is the development of a range of fiber-optic chemical sensors and systems for in situ use (Tebo, 1982; Seitz, 1984; Smutz, 1984–1985). Fiber-optic chemical sensors make use of either direct optical measurements down a fiber or one of many immobilized membranes or reagents at the fiber tip that reversibly or irreversibly bind with specific analytes, producing a response that can be sensed optically. A simple example is fiber-optic-guided fluorescence, which provides a direct measure of concentrations of polyaromic hydrocarbons and other compounds that fluoresce at the wavelength sent down the fiber (e.g., Inman et al., 1989, 1990). Ligands that fluoresce when bound to an analyte (such as a dissolved metal) can be either pumped to the fiber tip or immobilized on the fiber, providing a signal when the contaminant is encountered. A number of these sensors use biological coatings that can be selected for sensitivity to particular pollutants. To date, biosensors have been used primarily in the medical sciences, but they are also being used to monitor food quality and environmental conditions (Keeler, 1991; Schultz, 1991). Fiber optics are also used in light-addressable potentiometric sensors, which are solid-state devices sensitive to a variety of biochemical reactions (Hafeman et al., 1988). If suitably configured, biosensors may be particularly good for long-term monitoring of biological responses to selected contaminants, an area of special importance and considerable research difficulty.

In addition to applications dealing with a specific contaminant or reaction, fiber optics have been incorporated into a miniature spectrometer, permitting high-resolution measurements of fluorescence, absorbance, reflectance, and radiance in a variety of materials (Ocean Optics, 1996). Although demonstrated in a number of industrial applications, the system has not yet been tested in the marine environment.

These new systems represent a promising beginning, but there is still a clear need for the identification, development, and demonstration of new and improved chemical sensors (both remote and in situ) for measuring contaminant concentrations in marine sediments. The availability of such sensors would contribute significantly to the development of improved management protocols for contaminated sediment sites.

Survey Design: Numerical Simulation Methods

Designs of sediment sampling strategies, and identification of optimum remediation methods, increasingly rely on computer-based numerical models. In concept, these simulations have the potential to highlight the key parameters requiring field measurement, assess the sensitivity of the study area to aperiodic storms, and forecast changes resulting from a specified remediation scheme. These models fall into four general categories: hydrodynamic, sediment and chemical transport, biological toxicity, and ecosystem response. Each has a characteristic range of strengths and weaknesses.

Numerical modeling of coastal and estuarine hydrodynamics has been a subject of interest for more than 30 years (Ward and Espey, 1971; Fischer, 1981). In combination with increasingly sophisticated field observations (e.g., acoustic Doppler current meters [Brumley et al., 1991]) and advances in computing power and capability, hydrodynamic models provide reasonably accurate simulations of a variety of flow conditions. Many of these models are available through commercial vendors. To produce reliable results, however, the models need to be used by professionals who are familiar with coastal dynamics and numerical methods; they also require a comprehensive set of field data for calibration and verification.[4] Two-dimensional representations of the flow field are routinely available,

---

[4] Predictive modeling requires a research strategy, not simply a monitoring approach.

with improvements in computational efficiency favoring increasing availability and the development of three-dimensional models. Such models can accurately evaluate the effects of extreme events, such as floods and hurricanes, on local hydrodynamics, and make quantitative estimates of any effects of remediation projects, such as changes in water depth as a result of dredging or the placement of sediment caps to isolate contaminated deposits.

Although the modeling of coastal hydrodynamics is relatively advanced, efforts to couple the resultant flow field to the underlying sediment column have been hampered by the complexity of the factors governing the erosion and entrainment of fine-grained sediment. As described above, the structure and strength of a sediment deposit can be expected to display significant spatial and temporal variability. To date, this variability has precluded the derivation of a generally applicable transport algorithm. As a result, accurate simulations of sediment and chemical transport require that site-specific formulations be developed for each project. As a minimum, measurements to assess sediment erodibility under varying boundary-shear stress conditions and estimates of particulate settling velocities over a range of concentrations and water temperatures are required.

Several models, such as the USACE-developed TABS-2 (Thomas and McAnally, 1985) and HEC-6 (Hydrologic Engineering Center, 1993), that can predict coarse-sediment transport are available. Both TABS-2 and HEC-6 have been used for contaminated sediment sites. Similar models for the transport of cohesive sediments would be useful for estimating contaminant dispersion while remediation alternatives are being explored or even implemented. Unfortunately, transport and contaminant partitioning are much more complicated in cohesive sediments than in coarse sediments, and such models are not readily available.

The uncertainties associated with the numerical predictions of fine-grained sediment transport complicate the quantification of the exposure of resident biota to sediment-associated contaminants. Suspended sediment concentrations, resuspension, and deposition rates represent primary input data for the commonly used numerical physical-chemical fate models (Thomann and Mueller, 1987). Inaccuracies in these data affect the estimates of contaminant concentration governing biotic exposure. In turn, exposure data typically serve as essential input for numerical models of species toxicity and subsequent ecosystem response. The lack of accurate data, combined with the limited understanding of the effects of an assemblage of contaminants acting individually and synergistically on selected species, significantly limits the accuracy of numerical models of toxic response. Ecosystem models are limited further by difficulties inherent in predicting system responses to the addition or removal of particular classes of contaminants, factors that govern the partitioning of contaminants within the food chain and prey-predator contaminant transfers. Considerable research will be needed before numerical models of species toxicity and ecosystem response can be widely used.

SUMMARY

Source control is advisable in all contaminated sediment management projects. There are, however, impediments, including the difficulty of identifying certain sources of contamination. Even if sources cannot be controlled, remediation efforts may still be warranted. At the same time, regulatory steps could be taken to improve source control. For example, the EPA and the states could consider the impact on sediment quality downstream in setting TMDLs for waterway segments and in developing load allocations for contaminant sources. Site assessment represents an essential element in the design and implementation of every contaminated sediment management plan. Carefully conducted site assessments can accurately define the nature and extent of contamination and facilitate effective remediation, thereby maximizing overall project cost effectiveness. Given the complexity of the coastal and estuarine environment and the number of factors affecting contamination, a well-structured and systematic approach to site assessment is needed.

The methods and procedures associated with detailed field surveys are in an early stage of development. Selected field tools are available but have not been widely used. As dredging methods and practices improve and pressure for cost reduction increases, the demand for increased survey resolution and more accurate short- and long-term predictions will force the development of improved survey methods and complementary numerical models.

The cost effectiveness of site assessment could be enhanced by the continued identification, development, and demonstration of innovative survey approaches. Acoustic profiling, if the state of the art evolves as some believe it will, has the potential to reduce the costs of site characterization and, perhaps, the costs of evaluating and monitoring capped sites. However, additional fundamental research is needed on the acoustic properties of marine sediments to determine if acoustic profiling can accurately define areas of contamination. In addition, chemical sensors (both remote and in situ) for measuring contaminant concentrations in marine sediments need to be developed and tested.

Site evaluation and assessment of remedial alternatives require conceptual, analytical, and numerical models that can predict hydrodynamics and contaminant transport, transformation, and biological effects. At this point, however, there is no fundamental understanding of cohesive sediments transport and the effects of contaminants on ecosystems. Models that enable comparisons between environmental effects can be used to gain some insights on site assessment.

REFERENCES

Apitz, S.E., L.M. Borbridge, K. Bracchi, and S.H. Lieberman. 1993. The fluorescent response of fuels in soils: Insights into fuel-soil interactions. Pp. 139–147 in International Conference on Monitoring of Toxic Chemicals and Biomarkers. K. Cammann and T. Vo-Dinh, eds. SPIE Proceedings, vol. 1716. Bellingham, Washington: Society of Photo-Optical Instrumentation Engineers.

Bohlen, W.F. 1993. Fine grained sediment transport in Long Island Sound: Transport modeling considerations. Pp. 67–72 in Long Island Sound Research Conference Proceedings. M. van Patten, ed. Connecticut Sea Grant Program, Publication No. CT-SG-93-03. Groton: University of Connecticut.

Brumley, B.H., R.G. Cabrera, K.L. Deines, and E.A. Terry. 1991. Performance of a broad band acoustic Doppler current profiler. IEEE Journal of Oceanic Engineering 16(4):402–407.

Caulfield, D.D., A. Ostaszewski, and J. Filkins. 1995. Precision Digital Hydroacoustic Sediment Characterization/Analysis in the Trenton Channel of the Detroit River. Paper presented at the 1995 Conference on Great Lakes Research held May 28–June 1, 1995. East Lansing: Michigan State University.

Environmental Protection Agency (EPA). 1994. EPA's Contaminated Sediments Management Strategy. Draft. EPA 823-R-94-001. EPA Office of Water. Washington, D.C.: EPA.

EPA and USACE. 1991. Evaluation of Dredged Material Proposed for Ocean Disposal. Testing Manual. EPA 503/8/91/001. Washington, D.C.: EPA Office of Water and USACE.

Fischer, H.B., ed. 1981. Transport Models for Inland and Coastal Waters. Proceedings of a Symposium on Predictive Ability. New York: Academic Press.

Gibbs, R.J. 1973. Mechanisms of trace metal transport in rivers. Science 180:71–73.

Hafeman, D.G., J.W. Parce, and H.M. McConnell. 1988. Light addressable potentiometric sensors for biochemical systems. Science 240:1182–1185.

Hayter, E.J. 1989. Estuarial sediment bed model. Pp. 326–359 in Estuarine Cohesive Sediment Dynamics. A.J. Mehta, ed. Lecture Notes on Coastal and Estuarine Studies, no. 14. New York: Springer-Verlag.

Hydrologic Engineering Center (HEC). 1993. HEC-6. Scour and Deposition of Rivers and Estuaries. User Manual. CPD-6. Davis, California: U.S. Army Corps of Engineers, HEC.

Inman, S.M., E.J. Stromvall, and S.H. Lieberman. 1989. Pressurized membrane indicator system for fluorogenic-based fiber optic chemical sensors. Analytica Chimica Acta 217:249–262.

Inman, S.M., P. Thibado, G.A. Theriault, and S.H. Lieberman. 1990. Development of a pulsed-laser, fiber optic based fluorimeter: Determination of fluorescence decay times of polycyclic aromatic hydrocarbons in sea water. Analytica Chimica Acta 239:45–51.

Keeler, R. 1991. Biosensor applications are springing up all over. R&D Magazine (May):62–66.

Lieberman, S.H., G.A. Theriault, S.S. Cooper, P.G. Malone, R.S. Olsen, and P.W. Lurk. 1991. Rapid, subsurface, in-situ field screening of petroleum hydrocarbon contamination using laser induced fluorescence over optical fibers. P. 9 in Second International Symposium—Field Screening Methods for Hazardous Wastes and Toxic Chemicals held February 12–14, 1991 in Las Vegas, Nevada.

McGee, R.E, R.F. Ballard, Jr., and D.D. Caulfield. 1995. A Technique to Assess the Characteristics of Bottom and Subbottom Marine Sediments. Technical Report DRP-95-3. Vicksburg, Mississippi: U.S. Army Engineer Waterways Experiment Station.

Moore, J.N., E.J. Brook, and C. Johns. 1989. Grain size partitioning of metals in contaminated coarse-grained river flood plain sediment, Clark Fork River, Montana, USA. Environmental Geology and Water Science 14(2):107–115.

National Research Council (NRC). 1987. Sedimentation Control to Reduce Maintenance Dredging of Navigational Facilities in Estuaries: Report and Symposium Proceedings. Washington, D.C.: National Academy Press.

Ocean Optics. 1996. The World's First Miniature Spectrometer. Technical brief and product description. Ocean Optics, Inc., 1104 Pinehurst Road, Dunedin, Florida 34698.

Ross, M.A., and A.J. Mehta. 1989. On the mechanics of lutoclines and fluid mud. Journal of Coastal Research (Special Issue)5:51–61.

Schultz, J.S. 1991. Biosensors. Scientific American (August):64–69.

Seitz, W.R. 1984. Chemical sensors based on fiber optics. Analytical Chemistry 56(1):16–34.

Smutz, M. 1984–1985. Fiber optics at sea. Ocean Science and Engineering 9(4):447–456.

Tebo, A.R. 1982. Sensing with optical fibers: An emerging technology. Pp. 1655–1671 in Proceedings of the Instrument Society of America Annual Meeting. Philadelphia: Cahners Publishing Company.

Thomann, R.V., and J.A. Mueller. 1987. Principles of Surface Water Quality Modeling and Control. New York: Harper & Row.

Thomas, W.A., and W.H. McAnally, Jr. 1985. Open Channel Flow and Sedimentation. TABS-2. User Manual (rev. 1990). HL-85-1. Vicksburg, Mississippi: U.S. Army Engineer Waterways Experiment Station.

U.S. Army Corps of Engineers (USACE). 1995. Micro-Survey Acoustic Core and Physical Core Inter-Relations with Spacial Variation. Washington, D.C.: USACE.

Ward, G.H., and W. Espey. 1971. Estuarine Modeling: An Assessment of Capabilities and Limitations for Resource Management and Pollution Control. Project 16070DZV. Water Quality Office. Washington, D.C.: EPA.

Ward, M.D., and D.A. Buttry. 1990. In situ interfacial mass detection with piezoelectric transducers. Science 249:1000–1007.

5

Interim and Long-Term Technologies and Controls

INTRODUCTION

When characterization of a site determines that contaminated sediment poses unacceptable risks to humans and/or ecosystems, the next step is the evaluation and selection of control measures. This chapter assesses the state of practice and the research and development (R&D) needs for both interim controls, which can be used to reduce high-risk levels quickly, and engineered technologies for longer term, more complete remediation. Costs, which often dictate the selection of technologies, are examined as well.

Numerous technologies and practices are available for managing contaminated sediments (NRC, 1989; Sukol and McNelly, 1990; EPA, 1991, 1993a,b), but few have been tested in marine environments. Although considerable experience with contaminated sediments in fresh water has been accumulated, and some of it may apply to marine systems, such extensions should be approached with caution. The high salt concentration in marine waters influences the surface chemistry of clays, their ion-exchange capacity for metals, and the resulting physical structure of the sediment. More important, perhaps, is the influence of high salt concentrations, particularly sulfate, on microbial processes.[1] Applicability to marine sediments is just one of many considerations in selecting a technology.

---

[1] Sulfate can be reduced to sulfides in organic-rich sediment, leading to the precipitation of metal contaminants. High sulfate concentrations can prevent methane formation, which takes place in organic-rich freshwater sediments. Because organic contaminant concentrations depend on whether conditions are methanogenic or sulfate reducing, freshwater and marine sediments are expected to undergo different intrinsic and engineered rates of transformation.

Harbor managers, state and federal authorities, city mayors, industrial plant managers, and military base commanders are often overwhelmed by the complexities of the technical issues as well as questions about costs, benefits, and the potential for hazard reduction, among other factors. This chapter attempts to sort through these issues to provide constructive guidance (see Box 5-1).

Much of the experience managing contaminated sediments, and hence the basis for much of the analysis in this chapter, comes from the Great Lakes, where the search for solutions began more than 20 years ago. A very high proportion of material dredged from the Great Lakes is contaminated, and open-water disposal became impossible by the early 1970s. Therefore, technology development and community-based debate and selection mechanisms are generally at later stages of development there than on the other coasts. The 1987 Amendments to the CWA authorized the EPA, in conjunction with other federal agencies, to conduct a five-year study of treatment processes for toxic pollutants in Great Lakes sediments. The resulting assessment and remediation of contaminated sediments (ARCS) research and planning program has provided much of the data on remediation technologies. The program and the overall results to date have been summarized by Garbaciak (1994) and EPA (1994a,b). Although the committee relied heavily on reports generated by the ARCS program, it must be emphasized that this data comes from freshwater systems. Furthermore, the reports are not readily available, have not been peer reviewed, and are based partly on anecdotal information.

BOX 5-1
Importance of Cost in Technology Assessment

Remediation technologies are costly, with costs escalating based on the number of steps. The cost of treatment, for example, is in addition to the costs of dredging and sediment placement or reuse.

The most effective technologies for eliminating contamination—that is, treatment or decontamination technologies—are the most difficult to implement, the most equipment intensive, and usually cost the most.

Although the costs of many technologies can be estimated, comparative data on costs of methods actually used in the field are limited and unreliable.

The effectiveness of remediation technologies in reducing risk has not been measured, so cost effectiveness can only be estimated.

The committee has emphasized throughout this report that a risk-based approach to the management of contaminated sediments is both logical and essential. Currently, however, controls and technologies are not assessed with regard to their risk reduction capability. The end-points now used (see Chapter 2) are not intended to determine whether remediation technologies actually meet the project goals. But post-project evaluations are conducted in some cases. For example, now that the Superfund site at Waukegan Harbor has been cleaned up to the standards set for the project (the removal of PCB concentrations above 50 parts per million [ppm]), the EPA plans to determine whether fish in the area are still contaminated (S. Garbaciak, EPA, personal communication to Marine Board staff, November 30, 1995). But so few contaminated sediment sites have been cleaned up that there is no consistent standard for post-project evaluations (S. Garbaciak, EPA, personal communication to Marine Board staff, November 30, 1995).

The overall goal in the remediation of contaminated sediments, therefore, remains the removal or isolation of contamination to meet human and ecosystem exposure limits. Achieving this goal at an affordable cost requires a systems approach to the evaluation of possible solutions, including natural recovery and other in situ approaches; sediment removal and transportation technologies; and ex situ controls, including treatment or decontamination. However, the range of choices in any given situation is limited by site conditions. High-unit-cost treatments are precluded, for example, for large volumes of sediments with relatively low levels of contamination. Similarly, the selection and sequence of ex situ treatment technologies are constrained by the characteristics of marine sediments, which, as long-term integrators of contaminants in aquatic environments, typically contain a complex matrix of organic and inorganic compounds that are both nonvolatile and relatively insoluble in water. If risks are high enough to be judged imminently hazardous, then interim controls must be used to reduce risk levels quickly.

Interim and long-term control technologies are a subsystem of the overall remediation system. Figure 5-1 is a schematic diagram of this subsystem and the various components that must be considered. The four major sections in this chapter address the four components of the subsystem: interim control technologies; in situ management technologies; sediment removal and transportation; and ex situ management.

The first section examines interim controls, including administrative and technology-based measures, which may be used to reduce imminent hazards. The second major section deals with in situ management, including natural recovery processes that reduce contaminant bioavailability through either destruction or isolation; in-place contaminant isolation by capping; and active treatment through thermal, chemical, or biological processes. The third section addresses sediment removal and transportation by dredges, pipelines, and barges for environmental, as opposed to navigation, purposes. (Land-based transportation by truck or rail is not addressed in this report.) The assessment focuses on criteria for selecting

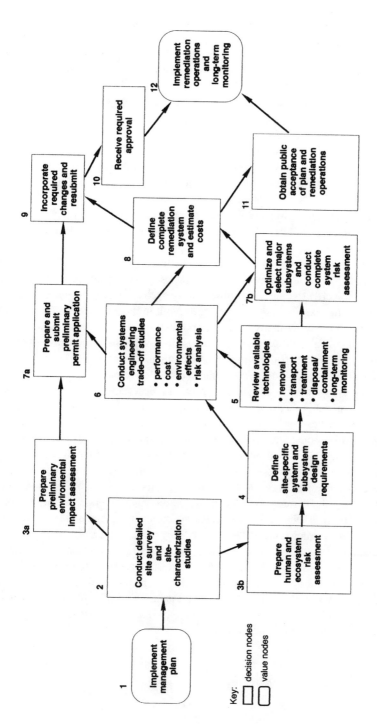

FIGURE 5-1 Process of defining a remediation system. Note: See Box 5-2 for details.

equipment and on environmental impact. The last section assesses ex situ treatment and containment management, which encompasses dozens of technologies.

The chapter concludes with three sections that integrate information on the four components. One section examines how the performance of technologies and controls is evaluated through monitoring, estimates of cost-effectiveness, and other activities. Another section summarizes needs for R&D, testing, and demonstrations. The final section presents a qualitative comparison and overall assessment of the various categories of technology.

The use of remediation technologies and controls in the management of contaminated marine sediments is still emerging. For the most part, the field has been dominated by tools developed for navigation dredging, and few full-scale treatment systems have been implemented. Therefore, the committee's analysis focuses on the general classes of treatment technologies that are applicable to treating contaminants found in sediments. The discussion is not very detailed, and the cost estimates are uncertain.

Technical developments worldwide are considered. All technologies are examined with respect to scientific and engineering feasibility, practicality, cost, efficiency, and effectiveness. Key attributes of each technology are noted in summary tables; the text does not reiterate each point in the tables but addresses only those issues that require analysis. It is important to note that performance can be evaluated only in a qualitative sense, because the available data on cost and effectiveness are inadequate for making reliable comparisons of technologies based on cost effectiveness or any other meaningful quantitative basis.

To achieve optimal results, decision makers must understand the role of technology assessment in the overall remediation system, which includes the elements discussed in earlier chapters, regulatory issues, stakeholder interests, and site-specific considerations. A simplified description of the remediation system is shown in Figure 5-2 and described briefly in Box 5-2. The system includes many of the elements found in the conceptual management approach presented in Chapter 2 (Figure 2-1). However, in the present context, the focus is on defining, integrating, and optimizing the various components of the remediation system. Only the most significant tasks are shown; the physical orientation among the tasks is based on the relative timing between tasks and the dependency on the completion of earlier tasks. The direction of data flow between tasks, both with respect to input data and the output of results, is shown by the arrows. For example, task 8 requires input from tasks 6 and 7. When it has been received and task 8 has been completed, the results of task 8 are used in tasks 11 and 12 as input data. The timing of the management schedule and technical risks affects costs directly and is an important consideration in the design of the remediation system.

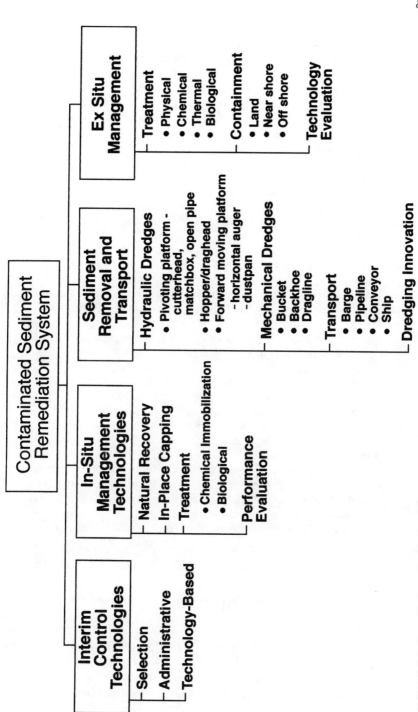

FIGURE 5-2 Remediation technologies subsystem structure.

> **BOX 5-2**
> **Process of Defining a Remediation System**
>
> The discipline and structured thought process inherent in systems engineering provides a logical approach to the development of acceptable and workable strategies for managing contaminated sediments. Before technologies and controls can be evaluated, the system boundaries must be defined (see Figure 5-1). In addition to establishing the physical boundary for the horizontal and vertical extent of contamination, the contaminants of concern, political institutions, applicable laws and regulations, regulatory bodies, stakeholder interests, planning-time horizon, and desirable end-points must also be bound. Without boundaries, the extent of the system is poorly defined, and reaching a near-optimum solution will be difficut.
>
> The various elements of the system must be determined. Defining the geographical extent of contaminated marine sediments (task 2) presents great difficulties. In many engineering processes, the physical boundary has controllable, or at least measurable, inputs and outputs. In the case of contaminated sediments, the boundary is less defined, and there is exchange of water, sediment, air, and aquatic organisms. Legal and regulatory constraints must also be recognized. Environmental laws and regulations at the local, regional, state, federal, and international levels constrain the management of contaminated sediments through environmental impact assessments and the permitting process (tasks 3 and 6). These limitations can delay the development of a solution and increase costs. Only at this stage can appropriate technologies be assessed (task 5).
>
> Once the objective functions have been quantified along with constraints, the optimal solution can be studied. These studies (task 6) address the interrelationships of the subsystems, considering performance, costs, and environmental effects. These studies permit definition of the optimal approach (task 7) to the selection of the appropriate removal, transport, treatment, and disposal subsystems of an integrated total system, designed within the available and proven component technologies. Other elements of the process include public acceptance of the proposed remediation plan (task 11).

INTERIM CONTROLS

A previous report by the NRC (1989) found that sediment contamination issues at Superfund sites were often not addressed effectively because of the time lapse between the identification of the problem and the initiation of remedial action. In the dynamic underwater environment, a long wait often means that the contamination has spread, making it much more difficult and costly to clean up than it would have been when it was concentrated in a small area. The 1989 report, therefore, recognized the value of using interim control measures soon after discovery of the problem to prevent the situation from deteriorating or to avert excessive damage over the prolonged period required to choose a long-term course of action and secure all necessary regulatory approvals.

For purposes of the present report, interim control technologies are defined as temporary measures that can be implemented quickly to meet an immediate need to control exposure to contaminants and reduce risk to humans and the environment. It is appropriate to consider interim measures in all cases where an imminent hazard has been identified by risk analysis (discussed in Chapter 2) and reasoned judgment.[2] Permanent solutions typically take 3 to 15 years to implement, according to the committee's case histories (see Table 1-1). By definition, interim controls must be less expensive than long-term controls and must be suited to faster implementation. Interim controls include a broad spectrum of administrative and technology-based approaches based principally on isolation or avoidance techniques. Controls considered by the committee range from issuing public warnings or health advisories to constructing barriers blocking access to contaminated areas by humans or other biota. Slow processes, such as natural recovery and bioremediation, are not included in this category (these approaches are examined as long-term solutions).

Experience with interim controls has been limited. The committee identified only a handful of cases in which such measures have been used, just two of which involved technology-based control. Nevertheless, there is some evidence that these measures are at least partly effective in the short term, and, equally important, they may offer the only hope of rapid risk reduction in highly contaminated areas. Indeed, interim controls are likely to be used more in the future because the costs of treating large volumes of sediment in "permanent" ways are generally very high. Interim controls require special attention in the planning process, however, because the importance of quickly reducing exposures and controlling the scale of the problem are often overlooked in the rush to find a more permanent response. The effective use of interim controls could be enhanced by monitoring and evaluating their effectiveness where they are being used.

Selection of Interim Controls

A decision to proceed with interim controls can be made at any point in the decision process after preliminary site data have been obtained. But inexpensive, fast-acting methods cannot be expected to provide permanent solutions. Decision makers who implement an interim strategy to address an imminent hazard must anticipate taking further, more elaborate action later to meet long-term cleanup criteria. It is possible, however, that interim actions or intervening events

---

[2] The focus is on when the hazard is identified, rather than when it developed (sediments tend to become contaminated slowly over time rather than suddenly).

may reduce the risk sufficiently to obviate the need for long-term measures. This phenomenon occurred at the James River, where commercial fisheries were closed in 1975 to reduce the health risk of Kepone contamination while decision makers considered permanent solutions (see Appendix C). Active remediation eventually was rejected as both too costly and environmentally unwise; in the meantime, Kepone manufacturing had been forbidden, and the contaminated sediments were covered over by clean sediments, a natural process that was effective enough to permit lifting of the fishing restrictions in 1988. Thus, the combination of an interim control and a passive, long-term solution (natural recovery) largely solved the problem (although maintenance dredging is still restricted, which is a problem because ships can navigate only at high tide). Post-project monitoring ensured that the risk was reduced indicating that no further action was necessary.

It is desirable, but not necessary, that interim control measures be compatible with, and possibly even complement, the ultimate solution. Interim measures that hamper long-term remediation can increase overall project costs. In some cases, the use of an interim control may reduce the overall project costs, but in the committee's view this is a side benefit rather than a selection criterion. Cost control is a consideration, however, in that an ill-conceived interim control might interfere with, or require expensive removal prior to, the implementation of a permanent solution. For example, a temporary sand cap might render dredging impractical, but extensive or armored capping is appropriate as a permanent solution and is not considered to be an interim control.

Administrative Interim Controls

Restrictions on catching or marketing high-risk fish and shellfish species can reduce the risks to human health in areas where unconfined contaminated sediments must remain in place for long periods of time (i.e., where natural restoration is planned or the selection and implementation of a remediation strategy drags on for years). Such restrictions can take various forms. In the James River case, commercial fisheries were shut down, which was a drastic step.

In areas frequented by recreational fishermen, other approaches may be necessary. For example, in the mid-1970s, the South Carolina Department of Health and Environmental Control and the EPA discovered that fish from certain areas of Lake Hartwell were contaminated with PCBs at levels above the Food and Drug Administration (FDA) tolerance limit of 5 milligrams per kilogram (mg/kg). To prevent or minimize exposure to fish with PCB contamination above a target risk level, the South Carolina Department of Health and Environmental Control issued a health advisory in 1976 warning the public against eating fish from the Seneca River arm of Lake Hartwell (EPA, 1994c; Hahnenberg, 1995). In 1984, the FDA lowered the PCB tolerance level to 2 mg/kg, and, as a result, the original health advisory was modified to specify that no fish taken in the highly contaminated areas should be eaten, nor should any fish larger than three pounds taken from the general area be eaten (Hahnenberg, 1995). Fishing was not prohibited, but signs warning against eating fish have been posted at most public boat launch areas and recreation areas at Lake Hartwell since 1987 (Hahnenberg, 1995). In addition, education programs designed to increase public awareness of the health advisory and methods of preparing and cooking fish were implemented to reduce further the quantity of contaminants consumed (see EPA, 1994c).

A health advisory is a temporary palliative because it obviously does nothing to minimize the exposure of, or risk to, fish-eating birds and mammals. But fishing restrictions can be left in place for years, even decades. The New Bedford Harbor Superfund site was closed to all fishing in 1979; in 1990, a number of studies culminated in a decision to remove and incinerate the sediments in hot spots (EPA, 1990). In some cases, fishing restrictions have been in place for so long that they have become de facto permanent solutions. For instance, PCB-contaminated fish and sediments were found in the upper Hudson River in the early 1970s. Health advisories against fish consumption from the lower river and a complete ban on fishing in the upper river have been in effect since the mid-1970s (Harkness et al., 1993).

Although complete bans on fishing can reduce risk to humans, the effectiveness of public advisories about contaminated sediments is an open question. The committee was unable to find enough information to document or analyze the risk reduction of either fishing bans or advisories. The compliance problems involved are illustrated by Belton et al. (1985) in a study that addressed a potential 60-fold increase in the risk of human cancer associated with the lifetime consumption of PCB-contaminated fish from the Hudson-Raritan estuary area. The effectiveness of public health advisories as risk reduction measures was evaluated by a careful, multidisciplinary study of recreational fishermen. Approximately 59 percent of those surveyed fished for the purpose of catching food. More than 50 percent of the respondents were aware of the warnings, and those who did not consume the fish generally were persuaded by a perception of unacceptable risks. But 31 percent of those who ate their catch did so despite believing it was contaminated. The researchers concluded that the broad-scale rejection of the health advisories was due to a combination of factors: the way the media were used, the nature and delivery of the health advisory, and personal predispositions that tended to reduce the credibility or usefulness of the communication.

Technology-Based Interim Controls

The committee could identify only two instances in which a technology-based interim control was implemented to control the dispersion of contaminated sediments. The use of technology-based measures may be impeded by concerns that, because of the cost associated with implementation or removal, they will narrow the choice of long-term solutions or become de facto, second-rate permanent solutions. There is also some question about how to monitor the effectiveness of interim controls. Nevertheless, in cases where a quick, inexpensive risk reduction is needed, a strong argument can be made for considering interim structural controls, preferably immediately after a high-risk site has been discovered.

Contaminated sediments can be covered with a layer of cleaner sediment or placed within a temporary containment structure, with the intention of removing them later for extensive or permanent treatment or disposal. This approach was demonstrated in 1995 when sediments were contained temporarily in Manistique Harbor in Michigan to prevent the resuspension and transport of PCB-contaminated sediments into Lake Michigan (Hahnenberg, 1995). A high-density polyethylene plastic liner (110 feet by 240 feet) was placed over the hot spot with the highest surficial PCB concentration at a cost of approximately $300,000. One-way gas valves and more than 40 2,000-pound concrete blocks were installed to keep the liner in place. This measure was used until a permanent cap could be installed. The effectiveness of the temporary cap was evaluated by monitoring the liner placement to ensure that the hot spot remained covered. It is not known, however, if the cap actually reduced the risk posed by the PCB contamination.

The second case of a structural interim control known to the committee was at New Bedford Harbor, where limited dredging of a hot spot was combined with the temporary storage of sediments for later treatment (Otis, 1994). The dredging of hot spots is analogous to short-term Superfund "removal"[3] prior to more extensive "remedial response." When contaminated sediments must be moved out of a navigation channel, it may be cost effective to remove and store the sediments until final treatment and disposal methods can be selected. In such cases, removal and storage not only reduce the immediate risk but also serve as necessary components of the ultimate solution. Sediments can be stored, for example, in a CDF. Although CDFs are generally not used in rapid response to imminent hazards, it is possible to recover and reuse CDFs by following a series of steps, including solids separation, dewatering, and removal of the sediments to a permanent disposal site or for beneficial use. In the New Bedford case, a CDF was to be used for both a pilot study and the hot-spot remediation. Eventually, it was capped (Otis, 1994). Management guidelines are available for the reuse of dredged material disposal areas (Montgomery et al., 1978). Dredged contaminated sediments have also been placed temporarily in multicelled settling basins for treatment (as in the Marathon Battery case history) and in confined aquatic sites (as in the Port of Tacoma case history).

Another interim approach involves the installation of sediment traps and by-pass systems, which can redirect the deposition of new contaminated sediments to a controlled location or can isolate "clean" natural sediments from highly

---

[3] The term "removal" as used by Superfund is not necessarily confined to physical excavation. The term refers to a broad array of "emergency" response measures, which require less time and money to implement than longer-term, more permanent "remedial response" measures.

contaminated sites or sources. In the context of navigation dredging, these measures can help control the accumulation of contaminants in the intervals between routine maintenance dredging or during the lengthy process required to secure authorization and funding for new-construction dredging. Sediment traps were used on a long-term basis in Indiana Harbor to control sediments entering a contaminated zone and thereby reduce the volume that had to be dredged. Construction of a sediment trap has been proposed for the same reason in Michigan City, Indiana (Miller, 1995). The use of sediment traps downstream from a hot spot could be useful, on an interim basis, to keep contamination from spreading.

TECHNOLOGIES FOR IN SITU MANAGEMENT

In situ management involves one or more processes that do not require removing sediment from its original location. The contaminants are either destroyed in place, isolated, or immobilized to prevent significant releases into the ecosystem. In situ management includes natural recovery, in-place capping, and in situ chemical and biological treatment. In North America, these practices have been used at fewer than 20 sites. These processes may not be feasible in navigation channels, which require periodic dredging. In situ management processes require a commitment to long-term monitoring.

Natural Recovery

Natural recovery involves leaving the contaminated sediments in place and allowing the ongoing aquatic processes to contain, destroy, or otherwise reduce the bioavailability of the contaminants. Although no action is required to initiate or continue the process, natural recovery is considered the result of a deliberate, thoughtful decision. The same process may occur by default or as an interim approach at Superfund and other sites when cleanup is delayed by legal, technological, economic, or other barriers. Natural recovery is a viable approach if the contaminants are being buried by cleaner sediments or if ongoing processes destroy the contaminants so that contaminant transport into the overlying water column is minimal and decreases with time. Some natural recovery processes are obviously very effective, as has been shown by profiles of contaminant concentrations preserved in Canadian sediment beds since the 1940s (Wong et al., 1995).

Natural recovery has been a strategy of choice at two sites, including the James River in Virginia (Huggett and Bender, 1980), where natural sedimentation buried sediment contaminated by Kepone (see Appendix C), and Lake Hartwell, South Carolina (Hahnenberg, 1995). In general, natural recovery is not considered a deliberate choice but is viewed as the "no action" alternative in the context of the National Environmental Policy Act (NEPA) of 1969 (P.L. 91-190), which requires a complete assessment of all alternatives to proposed federal actions. Natural recovery is always a possibility if there is no need to dredge or otherwise disturb the site for the maintenance of navigation channels or for port development. A major advantage of natural recovery is low cost; the primary expenses are associated with the initial evaluation, the long-term monitoring, and indirect costs, such as the loss of commercial or recreational uses of the area. Summaries of the costs and other considerations are included in Table 5-1.

Among the limitations of natural recovery are that burial occurs only in depositional areas, and even these areas can be subject to erosion from anthropogenic processes or severe storms. A PCB hot spot identified in Saginaw Bay in 1989 was dispersed by a major storm in 1990, an unfortunate circumstance given that the contamination is now distributed throughout the bay and, although reduced in concentration, cannot be removed (S. Garbaciak, EPA, personal communication to Marine Board staff, December 1, 1995).

Other disadvantages are that the science of natural recovery is poorly understood. The in-bed processes that govern chemical containment or destruction, for example, are not well understood, and measurement can be difficult because of the complexity and variability of natural processes. To determine in situ chemical fluxes from the sediment bed to the water column, not only must the diffusion flux be measured at the sediment-water interface, but estimates must also be made of sediment erosion, advective flows within the sediment, and the dynamics of sediment reworking by the complete benthic community. At most sites, the relative contributions of these mechanisms are not known. The current lack of the capability of quantifying chemical movements accurately precludes a definite determination of the risk posed at a site being considered for remediation by natural recovery. It is seldom known, for example, the percentages of in-bed contaminants that undergo intrinsic degradation, are buried deep within the bed, are released to the water column by passive processes, such as diffusion or active biological processes, are extracted by organisms migrating and feeding, or are moved by erosion and resuspension.

The monitoring strategy at a site undergoing natural recovery must test the claim that the numerous relevant processes are indeed operating to isolate or eliminate the offending chemicals. If site conditions indicate the need for intervention, then a more active approach can be applied that better controls the risk to humans and the ecosystem. A sound strategy for monitoring natural recovery would include measurements of processes that can be measured, such as sediment accumulation rates, contaminant levels in the sediment by depth, bioaccumulation by benthic organisms, and the migration or harvesting of contaminated organisms. It would also be useful to know the chemical release rates from the bed and in-bed chemical transformation rates, but these processes are difficult, if not impossible, to measure. The monitoring strategy for natural recovery must be more carefully planned and implemented than for other technologies because it is assumed that there will be some chemical release—although at a low, and therefore tolerable, rate.

TABLE 5-1 Natural Recovery

State of Practice (system maturity, known pilot studies, etc.)	Applicability	Advantages/Effectiveness	Limitations	Research Needs
Selected for James River Kepone contamination and considered at Port of Tacoma site.	(a) Bed is stable or depositional; (b) chemical release rates are low; (c) interim controls can maintain safety to health and environment; (d) contamination level at active surface is low, but areal extent is large; (e) most of the contamination is below the bioturbed zone; (f) contaminants are underlain by low-permeability strata; (g) site is not subject to dredging or other disturbance; (h) source of contamination has been abated.	(a) There may be less environmental risk to await natural capping than to attempt sediment removal; (b) removal may cause physical harm to bottom communities as well as suspend and disperse contaminants; (c) cleanup cost may be prohibitive because of large area and low level of contamination; (d) low cost.	(a) Effectiveness of in-bed processes that govern chemical containment and/or destruction is poorly known; (b) bed remains subject to resuspension by storms or anthropogenic processes; (c) should only rarely be used in beds of flowing streams; (d) not appropriate if dredging is required or bulk quantities of chemicals, such as nonaqueous liquids or solids, are present.	(a) Develop scientific principles to describe the process of natural recovery; (b) based on a literature survey, document the success, failure, effectiveness, etc. of sites that have undergone natural recovery either by design or default; (c) develop accepted measuring protocols to determine in situ chemical flux from bed sediment to the overlying water column; (d) develop protocols for assessing the relative contribution of the five or more mechanisms for chemical release or movement from bed sediments.

Natural recovery is an attractive strategy. However, the scientific and engineering understanding needs to be improved through the development of a theoretical foundation to describe the process and verification by a comprehensive study of available data at sites where natural recovery has occurred. Long-term monitoring is needed to provide assurance that the process is effective. Identification of the critical issues requiring R&D at the laboratory and field scales could then be undertaken. The results of R&D would lead to the development of guidelines and criteria.

In-Place Capping

In-place capping is the controlled, accurate placement of a clean, isolating material cover, or cap, over contaminated sediments without relocating or causing a major disruption to the original bed. Caps usually consist of natural, granular materials, such as sand, although uncontaminated mud, geosynthetic materials, and armor stone have also been used. Capping is intended to stabilize the original bed against erosion and isolate the contaminants from contact with the benthic community, thereby reducing long-term environmental damage. Capping is an engineered procedure that can be used at appropriate sites, and its success depends on the careful design, construction, and long-term maintenance of the cap (Palermo, 1991a).

Capping is considered an appropriate measure for preventing benthic effects in the USACE dredging regulations (33 CFR §335 to §338; USACE and EPA, 1992) and is recognized by the International Convention on the Prevention of Marine Pollution by Dumping of Wastes and Other Matter (commonly known as the London Convention of 1972) as a management technique that rapidly renders unsuitable materials harmless (Edgar and Engler, 1984). A review of the literature conducted by Zeman et al. (1992) determined that at least 20 major capping projects have been conducted worldwide, including more than 10 in situ projects in North America. Evaluation of these projects has led to development of a preliminary understanding of the data, equipment, and procedures needed for successful capping. Palermo (1991a) presents a concise guide for all capping projects, and Shields and Montgomery (1984) provide an overview of engineering considerations for capping projects. Guidelines for in situ capping are in preparation (Palermo et al., in press).

Capping can be considered where discharges of contaminants have been halted substantially but natural recovery is too slow to solve the problem. Capping can also be considered where the costs and environmental effects of moving contaminated sediments are very high. However, capping may not be appropriate where the cap may be disrupted or scoured (e.g., from high-energy conditions, ice scouring, or heavy boat traffic) or where navigation dredging is a priority. Suitable capping materials need to be available in the requisite type and quantity to create the cap, and suitable hydraulic conditions must exist so the cap is not compromised. In addition, the original bed must be able to support the cap, and the capping material must be compatible with the existing aquatic environment. As with other in situ strategies, long-term monitoring is required to determine the effectiveness of capping.

The costs of complete capping projects have not been well documented, but numerous feasibility studies reporting cost estimates (Krahn, 1990; West Harbor Operable Unit, 1992) indicate that, except for natural recovery, capping is the least expensive in situ or ex situ remediation technology (also see Averett et al., 1990; EPA, 1994b). The reason for the low cost is that the contaminated sediments do not have to be moved or treated. If the cost of long-term monitoring is not included, then capping at some sites can be as inexpensive as open-water disposal, although more elaborate, costly capping technologies could be required at other sites. Table 5-2 summarizes key considerations with regard to capping.

Among the major benefits of in situ capping are that it eliminates the need to move the contaminated sediments and that it promotes the in situ isolation of the contaminants by significantly retarding their release to the benthic community. For example, an estimated 99 percent reduction in release rates of Aroclor-1254 and Aroclor-1242 (PCBs) to the overlying water column is predicted with a 0.45-m-thick sand cap in New Bedford Harbor (Thibodeaux et al., 1990). However, these calculations have not been verified, and they are based on the assumption that the cap has not been damaged by erosion or shipping. If the design of the cap is simple, capping material can be replaced, augmented, or repaired easily.

Capping may have some drawbacks, however. If caps are made of different materials than the ambient bottom sediment, they may alter the benthic community. Capping is not likely to be economical if the area of contamination is large. Capping is not suitable for use with highly contaminated bottom material consisting of organic sludges, hazardous solid waste, or other substances with characteristics different from the natural bottom sediment.

Another significant issue affecting the use of capping is the regulatory framework for sediment management. Under Superfund (§121[b]), a strong preference is given to treatments that "permanently and significantly reduce the . . . toxicity or mobility" of contaminants. Capping is not currently considered a permanent solution, even though it capitalizes on the natural tendency of contaminants to remain bound to sediments in low-energy sinks. The committee identified several ways to overcome this problem.

There is a precedent for viewing containment measures as the presumptive remedy for municipal landfills that have become Superfund sites. In the same way, it may be possible to secure a "preferred" remedy status for physical containment strategies, such as capping, by making them more permanent and geared more to reducing toxicity. For example, capping could be augmented by promoting in-place biodegradation, perhaps by injecting micronutrients or microorganisms (an untested approach that may be particularly useful for persistent

TABLE 5-2 In-Place Capping

State of Practice (system maturity, known pilot studies, etc.)	Applicability	Advantages/Effectiveness	Limitations	Research Needs
Less than 10 major in situ capping projects in North America have been completed (more than 20 worldwide). Reviews exist concerning (a) necessary data, equipment, and procedures; (b) engineering considerations; (c) guidelines for design of cap armor; and (d) predicting effectiveness of chemical containment.	(a) Contaminant sources have been substantially abated; (b) natural recovery is too slow; (c) costs and environmental effectiveness of relocation are too high; (d) suitable types and quantities of cap material are available; (e) hydrologic conditions will not compromise the cap; (f) cap can be supported by original bed; (g) appropriate for sites where excavation is problematic or removal efficiency is low.	(a) Eliminates need to remove contaminated sediments; (b) effective in containing contaminants by reducing bioaccessibility; (c) promotes in situ chemical or biological degradation; (d) maintains stable geochemical and geohydraulic conditions, minimizing contaminant release to surface water, groundwater, and air; (e) relatively easy to implement; (f) eliminates bioturbation and resuspension; (g) reduces contaminant release to water column; (h) easily replaced or repaired; (i) in shallow water, creates wetlands, dry lands, or reduces water column depth.	(a) Cap incompatible with bottom material can alter benthic community; (b) subject to erosion by strong currents and wave action; (c) subject to penetration/destruction by deep burrowing organisms; (d) destroys/changes benthic communities/ecological niches; (e) requires ongoing monitoring for cap integrity; (f) dilutes contaminants in original bed if subsequent removal/remediation is required.	(a) Analysis of data from existing and ongoing field demonstrations to support capping effectiveness; (b) controls for chemical release during bed placement and consolidation; (c) test to simulate and evaluate consequences of episodic mixing, such as anchor penetration, propeller wash, and/or mechanical penetration.

chemicals such as PCBs), or by adding activated carbon to the physical cap to adsorb certain contaminants. The ultimate solution would be to change Superfund regulations to allow and encourage capping where it is deemed appropriate.

The purpose of monitoring a capped site is to ensure that adequate cap thickness is maintained and that chemical penetrations into the clean cap material proceed at the projected design rate. Monitoring must also verify that biological penetrations from above are effectively limited by the cap. In essence, the philosophy behind the monitoring is that the cap is an engineered structure, similar to a bridge or a road, the performance of which must be verified and which must be maintained over time. If monitoring indicates thinning of the cap in some spots due to erosion, then fresh material or armoring would be required.

In sum, capping is one of the few accepted in situ techniques in use today, and the knowledge base for this technology is larger than for most other in situ technologies examined in this chapter. However, the knowledge base for capping is still incomplete because of a dearth of monitoring data. The precision of the cap placement could also be improved. Theoretical models and laboratory procedures are being developed that can be applied directly to the design and analysis of sediment caps. But monitoring methods must be developed for evaluating, on a case-by-case basis, the effectiveness of caps in preventing sediment erosion and minimizing the exposure of benthic organisms, measuring chemical fluxes into the overlying water column, and establishing that the level of risk is reduced to an acceptable level. Few data are available on long-term chemical fluxes through or out of caps; tidal and wave pumping and ship wakes are among the factors that may affect chemical fluxes. In addition, bottom profiling instruments are needed to verify cap thickness and to provide ongoing monitoring of cap integrity.

In Situ Treatment

In situ treatment involves adding unconfined chemicals or agents to the environment to immobilize or break down contaminants. In situ treatment poses numerous technical problems and has been used at very few contaminated sites, including several small sites in North America. Nevertheless, attention must be given to several in situ treatments, including immobilization, chemical treatment, and biological treatment.

Immobilization and Chemical Treatment

The goal of in situ immobilization is to isolate sediment contaminants from the benthic and aquatic ecosystem. The immobilization techniques considered most often are solidification and stabilization. The state of the art is summarized in Table 5-3. Solidification implies the conversion of sediments into a solid block with a structural integrity that physically binds the contaminants. Stabilization or chemical immobilization usually involves the addition of chemical reagents that

TABLE 5-3 Immobilization (solidification/stabilization)

State of Practice (system maturity, known pilot studies, etc.)	Applicability	Advantages/Effectiveness	Limitations	Research Needs
Manitowoc Harbor, Wisconsin, and Japan.	Limited	Not known	Not tested nor is in situ implementation likely	Extensive if technology is to be evaluated

TABLE 5-4 In Situ Chemical Treatment

State of Practice (system maturity, known pilot studies, etc.)	Applicability	Advantages/Effectiveness	Limitations	Research Needs
Unknown	Limited because of interference with other contaminants; possibly of mobilizing metals in the process of oxidizing organics.	Unknown	(a) Sediment would have to be isolated during mixing with reagents; (b) likely to bind with natural organic matter, oil, grease, and sulfide precipitates; (c) metals present in the sediments might be mobilized.	Extensive, if it is proposed, but probably not justified

reduce the solubility or mobility of the contaminants, with or without changing the physical characteristics of the treated material (EPA, 1989). An example of this treatment for industrial waste involves the addition of sulfides or elemental sulfur to promote the formation of metal sulfides, which have low solubility in water and therefore tend to form precipitates. However, because marine sediments of even moderate organic content are likely to be rich in sulfides, which naturally limit the mobility of metals, the addition of sulfides is not likely to be an appropriate treatment.

The in situ immobilization of sediments is likely to be based on the concepts of solidification and stabilization and to involve the addition of Portland cement, fly ash, or other binding agents to keep contaminated sediments in place and to reduce contaminant mobility. Immobilization reduces contamination through a combination of chemical bonding, encapsulation in a solid, reduction of permeability to reduce fluid flow, and reduction of pore space for diffusion. The applicability of the process to fine-grained sediments with a high water content has yet to be demonstrated. For this type of treatment to be efficient, the contaminated sediments need to be temporarily isolated to allow the mixing of reagents. Other potential problems could include inaccuracies in reagent placement, erosion, temperature increases during curing, and increases in sediment volume. Experience with immobilization techniques is not extensive enough to provide reliable estimates of the costs of large-scale treatments, their effectiveness, or possible toxic by-products.

Immobilization has been used on a small scale at Manitowoc Harbor in Wisconsin, where a cement and fly ash slurry was added to the sediment using a proprietary mixing tool and slurry injector (EPA, 1994b). The in situ mixing of cement with sediments for the primary purpose of enhancing compressive strength has not been proved or accepted for treatment of contaminated marine sediments in the United States (EPA, 1993a). Costs are estimated at \$15/yd^3 to \$160/yd^3 based on proposed applications (EPA, 1994b).

In situ chemical treatment involves the addition of chemical reagents to sediments to destroy organic contaminants. Theoretically, oxidants, such as ozone, hydrogen peroxide, and permanganate, could destroy PCBs and polyaromatic hydrocarbons. Chemical treatments would be difficult to implement because they require isolation during sediment mixing, and natural organic matter, oil and grease, and metal sulfide precipitates has very high oxygen demand. Furthermore, metals present in the sediments might be dissolved into the pore water after sulfide oxidation. Chemical dechlorination under ambient temperatures and typical water contents is not likely to occur or to be controllable.

Researchers at the Canadian National Water Research Institute have developed and demonstrated equipment capable of injecting chemical solutions into sediments at a controlled rate (EPA, 1994b). Chemical treatment of lake sediments to control eutrophication or to oxidize organic matter has also been demonstrated (EPA, 1994b). However, neither has been applied to the treatment of in situ contaminated marine sediments (see Table 5-4).

Other technologies can be mentioned here but quickly dismissed. In situ soil-water freezing on a permanent basis requires the presence of a refrigeration plant on site. Freezing by injection of molten sulfur, which has a melting point of 120°C, has the same limitations as in situ solidification (in addition to being unstable in marine systems because of its solubility and other reactions with dissolved salts). In situ vitrification has been demonstrated to isolate metals in soils, but high-water-content sediments with organic contaminants would require local site dewatering and vapor recovery.

Biological Treatment

In situ bioremediation[4] is used to hasten the natural restoration of the environment. The process involves fostering microbial biodegradation by providing needed but absent materials, such as oxygen, nutrients, or inoculants containing microbes known to be effective degraders of specific contaminants. The microbially mediated biodegradation of contaminants, both with and without intervention, has been observed at sites contaminated by a variety of organic compounds, such as crude and diesel oils; petroleum products; the aromatic hydrocarbons benzene, toluene, and xylene; PCBs; polyaromatic hydrocarbons; chlorinated phenolics; and many pesticides.

Numerous factors characteristics limit the use of biodegradative processes (see Table 5-5). The complexity of the sediment-water ecosystem, the difficulty of controlling the processes (physical, chemical, and biological) in the sediment, and the need to adjust environmental conditions for various stages of biodegradative processes limit the effectiveness of in situ bioremediation (EPA, 1994a). Although considerable research on in situ bioremediation has been carried out for a decade or more with soil systems, future R&D to overcome some of the difficulties and limitations with sediments may be costly. The best current alternative is in situ bioremediation using an engineered treatment system containing a portion of the bed sediment in cells, which allow reaction conditions to be controlled (see discussion of ex situ bioremediation later in this chapter).

In situ bioremediation was carried out on a rocky, petroleum-coated shoreline in 1989. A two-year study indicated that biodegradation of the oil could be stimulated by the addition of nitrate and phosphate, and that the rate of oil removal from beaches could be hastened (Pritchard and Costa, 1991; Bragg et al.,

---

[4] Bioremediation of contaminated sediments is defined by the EPA as "a managed or spontaneous process in which microbiological processes are used to degrade or transform contaminates to less toxic or nontoxic forms, thereby remedying or eliminating environmental contamination" (EPA, 1994a).

TABLE 5-5 In Situ Bioremediation

State of Practice (system maturity, known pilot studies, etc.)	Applicability	Advantages/Effectiveness	Limitations	Research Needs
(a) None documented for marine sediments; (b) examples from freshwater sediment are limited to special cases on pilot scale, e.g., chemical stimulation of dehalogenation (but no degradation) of PCBs in the Housatonic River, Connecticut; (c) stimulation of degradation with addition of active microbes in Hudson River.	(a) Contaminant is biologically available; (b) concentration of contaminant appropriate for bioactivity, e.g., sufficiently high to serve as substrate or not high enough to be toxic; (c) limited number or classes of contaminants that are biodegradable; less known for complex mixtures; (d) site is reasonably accessible for management and monitoring; (e) rapid solution is not required.	Based on experience from soil systems, it offers the potential for (a) complete degradation and elimination of organic contaminants; (b) reduced toxicity of sediment from partial biotransformation; (c) less materials handling, which can result in substantially lower costs; (d) no need for placement sites; (e) favorable public response and acceptability.	(a) Not a proven technology for sediments (freshwater or marine); (b) likely to require manipulation and disturbance of sediment; (c) can require containment which limits volume that is treatable; (d) can require long time periods, especially in temperate waters; (e) ineffective for low level contamination; (f) not applicable to areas of high turbulence or sheer; (g) not applicable for high molecular weight polyaromatic hydrocarbons.	(a) Fundamental understanding of biodegradation principles in marine environments; (b) bioavailability of sorbed contaminants and the effect of aging; (c) exploration of anaerobic degradation processes for the largely impacted near-shore anoxic sediments; (d) laboratory, pilot, and field demonstration of effectiveness for marine sediments; (e) interaction of physical, chemical, and microbiological processes on biodegradation, e.g., sediment composition, hydrodynamics; (f) analysis of cost effectiveness; (g) exploration of combining in situ bioremediation with capping.

1994). Biodegradation of the contamination occurred even without intervention, but the rate was slower.

In situ bioremediation technologies are used in land-based soils to degrade many, but not all, contaminants. With respect to marine sediments, however, bioremediation technologies are experimental. Soil and groundwater bioremediation technologies cannot be transferred directly to in situ marine sediments for a number of reasons. Three major barriers are noted here.

First, because little is known about the degradative potential of marine microbial consortia, it is not known how well lessons learned in land-based systems will translate to marine systems. It is clear, however, that the geochemistry and hydrogeology of marine sediments differ from those of land-based systems, and that these differences are likely to affect the behavior and fate of the contaminants. In other words, fundamental research is needed to address the microbial, geochemical, and hydrological issues affecting bioremediation processes. Even if such research were pursued, it would be unlikely to lead to useable technology soon. Experimental and bench-scale tests have yet to be translated to the pilot scale and demonstration level.

Second, the introduction of nutrients and an oxidant source (e.g., oxygen, iron, manganese, nitrate) to in-place contaminants is a major challenge in a marine environment. Although the hydrodynamics of some groundwater systems allow for the pumping of enriched waters through the aquifer to the contaminated site, this technique cannot be used with marine sediments in a harbor or bay because of dilution and the lack of containment. Proposed scenarios for treating sediments have not been demonstrated widely and can pose difficulties. A demonstration project at Hamilton Harbor, Ontario, involving the injection of calcium nitrate into sediments, achieved a 79 percent reduction in low-molecular-weight compounds but only a 25 percent reduction in polyaromatic hydrocarbons (EPA, 1994b). The reduction was attributed to biodegradation. One concern is that the use of rakes and other injection and mixing equipment may resuspend materials and cause adverse environmental effects. Another unproven scenario involves depositing nutrient-rich pellets onto the sediments and relying on benthic bioturbators to move the materials down into the sediments.

Third, unlike subsurface aquifers and soils, marine sediment biota are linked intimately to the benthic food chain. Hence, any augmentation intended to make contaminants more bioavailable to beneficial microbes may affect the complex food chain community in unknown ways.

Even if these hurdles can be overcome, conditions at most contaminated sediment sites pose additional challenges. Because marine sediments typically are contaminated with more than one class of toxic chemical, the selection of a treatment process is complicated, and the efficacy of the process will be lower than for the treatment of simpler wastes. When combined, contaminants such as toxic metals, polyaromatic hydrocarbons, and PCBs can have inhibitory effects on or can interact with each other. The combination of multiple contaminant classes also tends to rule out a single treatment technology. Selection of a treatment regime that includes in situ bioremediation may be precluded by elevated levels of toxic metals that could constrain microbial growth or by the limited biological and chemical availability of contaminants.

Pilot studies for the in situ biological treatment of sediments are limited to a few examples in which the sediment volumes were small and the contaminant composition limited. The committee was told of four projects, two involving PCBs and two involving polyaromatic hydrocarbons (Thoma, 1994).

The in situ bioremediation of PCBs has been carried out in freshwater sediments in the Housatonic River in Massachusetts and the Hudson River in New York (Harkness et al., 1993). The volumes involved were less than a few cubic yards; only small portions of the contaminated sites were studied. The Massachusetts field studies were carried out based on laboratory data indicating that bromobiphenyls stimulate anaerobic microbial attack on PCBs and that highly chlorinated congeners are dechlorinated to produce molecules with fewer chlorine atoms. The pilot studies were successful in showing that after 373 days, the concentration of highly chlorinated congeners (containing 6 to 9 chlorine atoms per molecule) declined from 68 to 18 percent of all PCB molecules, with a corresponding increase in the species with fewer chlorine atoms. Although total PCB levels did not change, the data suggest that toxicity was reduced because the most toxic congeners with "dioxin-like" properties were preferentially dechlorinated. Whether this result constitutes remediation depends on regulatory requirements. Given the current regulations, which are based on total PCB content, the novel capability of stimulating anaerobic PCB transformation may have limited practical use, and further treatment would be needed.

Field studies in the Hudson River showed that in situ aerobic biodegradation is limited by physical and chemical factors unrelated to the microbial community. These factors include, for example, the sorption of contaminants into the sediment matrix and the consequent reduction in contaminant biogeochemical and biological availability, oxygen and nutrient availability, mixing, and the survival of externally amended active organisms.

In sum, the in situ bioremediation of PCB-contaminated sediment has only recently been recognized as a potential alternative. Although the technology looks promising, given the current level of application and the regulatory focus on total PCBs, it is unclear whether in situ bioremediation can achieve the cleanup levels required at a reasonable cost. If additional nutrients are needed, the sheer volume of and contaminant mixtures of most marine sediments will present difficulties for handling and monitoring.

Available evidence suggests that PCB dechlorination and biodegradation occur more slowly in marine sediments than in land-based systems, but the in situ degradation rates of sediments have not been measured with any reliability. Furthermore, bioremediation rates would be affected by site-specific characteristics, such as sediment composition, hydrodynamics, pore water composition, and benthic biology. So far, there are no reliable cost data, although figures from the small experiments are available. The total cost of the field demonstration in the Hudson River was $2.5 million, excluding the costs of manpower, for the treatment of approximately 3 m^3 of sediment. Estimates of the cost of encapsulation technology range from $50 to $60/yd^3 of contaminated sediment. The inoculum cost alone was estimated to be $30 to $40/yd^3 (Harkness et al., 1993).

The feasibility of in situ biodegradation of polyaromatic hydrocarbons has been considered at the laboratory scale (Thoma, 1994). Initial experiments found that microbial degradation of polyaromatic hydrocarbons could be stimulated in harbor sediments, but the approach was difficult to monitor and the effectiveness could not be evaluated.[5] Research is now under way on ex situ processes for the removal of polyaromatic hydrocarbons and metals.

SEDIMENT REMOVAL AND TRANSPORT TECHNOLOGIES

In some cases, contaminated sediments must be moved for ex situ remediation or confinement. Efficient hydraulic and mechanical methods—dredges, pipelines, and barges—for removing and transporting sediments are available but may have to be modified to mitigate additional risks to the ecosystem and to facilitate remediation. This section discusses these modifications and how they can be made.

Environmental Dredging

Some dredging operations are primarily for environmental cleanup, and some are primarily for the improvement and maintenance of navigation facilities. This does not mean that routine dredging operations are not, or cannot be, environmentally friendly. However, navigation dredging, which is usually designed to remove large volumes of subaqueous sediments as efficiently as possible, is not addressed here. Environmental dredging, by contrast, is designed to remove contaminated sediments in such a way that the spread of contaminants to the surrounding environment is minimized.

Contracts for most dredging projects in the United States are based on the volume of sediment removed, and contractors bid in unit prices (in dollars per cubic yard) for the removal, transport, and placement of sediments, with lump-sum costs for the mobilization and demobilization of equipment. With few exceptions, the lowest qualified bidder is selected for the dredging job. This approach encourages the removal of as much material as possible as quickly as possible. If the sediment is "clean," this emphasis is appropriate.

---

[5] All shake flask experiments indicated that the polyaromatic hydrocarbons were reduced from 3,000 to 300 ppm in 60 days, regardless of the additions (Thoma, 1994). The limited data available thus far do not indicate that in situ processes can be accelerated by the addition of nutrients or microorganisms.

However, in a systems approach to environmental dredging, the dredging is often carefully integrated with subsequent treatment and disposal. If the lowest-cost, maximum-volume approach is used for removing contaminated sediments, then substantial amounts of water and uncontaminated sediments can be captured along with the contaminated portion, necessitating the handling of large volumes of sediment and increasing treatment and disposal costs. In 1995, navigation dredging costs ranged from less than $1/yd^3 (in situ volume) to little more than $5/yd^3 (more for small jobs or remote sites).

Because of the precautions necessary for environmental dredging, the cost would certainly be higher. But committee members experienced in dredging estimate that total costs for removal and transport would not exceed $15 to $20/yd^3. Even these costs are relatively low compared with the cost of many treatment processes, which can be more than $100/yd^3. In a systems approach, the cost of the treatment or placement method is a consideration in deciding how precise site assessment and dredging need to be. Adjustments in the dredging process to minimize the capture of water and uncontaminated sediments can reduce the costs of treatment and placement and hence reduce overall project costs. This section examines three topics: the criteria for equipment selection, the environmental risks associated with dredging, and recent dredging innovations.

Equipment Evaluation and Selection

Dredging contaminated sediments for cleanup involves many of the same considerations as dredging for navigation. The available guidelines on the selection of dredging equipment and the advantages and limitations of various types of dredges (USACE, 1993) are generally applicable to environmental dredging. Evaluation criteria for specific equipment can be found in other publications (NRC, 1989; Averett et al., 1990), which provide extensive discussions of available equipment and their operating characteristics.

When contaminants are not a concern, equipment is evaluated for its capability to operate under the anticipated site conditions, its compatibility with available sediment placement options, and costs. For example, hydraulic dredges, which employ centrifugal pumps to draw up sediment in a liquid slurry form and then transfer it to a pipeline, are generally used when large volumes of sediment must be removed, when the placement site is within pumping distance, and when the pipeline is not a major traffic obstacle. Mechanical dredges, which scoop up material with bucket-like equipment using mechanical force, are generally used to minimize sediment dispersion and to limit the effects of the dredge on sediment properties. These dredges are appropriate when sediment volume is relatively small or the placement site is not within pumping distance. Trailing hopper dredges (specialized hydraulic dredges), which operate from floating platforms that double as repositories for excavated sediments, are used primarily when site conditions include high waves, heavy traffic, and remote disposal areas. These site conditions often preclude the use of mechanical or hydraulic pipeline dredges, which are mounted on platforms that are less stable in waves than the platforms used with hopper dredges.

For environmental dredging, the equipment evaluation and selection process must be integrated with removal and transport technologies into an overall remediation plan. The precise removal of contaminated sediment, as well as the resuspension of sediment and the associated release of contaminants, are key concerns in the removal of contaminated sediments. Some dredges that can accomplish environmental objectives are already available. For example, a state-of-the-art backhoe dredge was designed specifically to remove creosote-contaminated sediments at the Bayou Bonfouca Superfund site in Louisiana. Significant elements of this dredging system include an array of position sensors installed on the backhoe arm and excavator; a computer-based monitoring system that enabled the operator to monitor turret rotation, arm angle, and bucket angle and depth relative to the vessel, and a topographic map of the bottom; and a slurry processing unit that greatly reduced the water content of the dredged material (Taylor, 1995). The dredge had a reach of up to 40 ft, adequate for many contaminated sediment sites.

The following sections discuss how to optimize an integrated remediation system, from sediment removal to the placement of residuals. Three evaluation criteria are discussed: site compatibility, precision and accuracy of removal, and the characteristics of delivered material.

Site Compatibility

The characteristics of, and access to, a site sometimes limit the kind of equipment that can be used in a specific project, unless equipment is specially adapted. Land-based equipment can be used to reach sediment within its operating radius or where the site can be dewatered and the sediment can support the equipment. Most projects require a floating dredge plant, which may be mounted permanently, mounted temporarily on a barge, or attached to a seaworthy vessel. Many contaminated sediments, however, are in backwater areas too shallow for some vessels, particularly during low tide. Access to these areas may also be restricted by low bridges and shallow channels, which limit the choice of vessels to those transportable by rail or truck.

Loading rates and material characteristics must match the capabilities of the treatment facilities. Therefore, removal and transport equipment must be capable of delivering sediments in the appropriate form. The automated system used in the Bayou Bonfouca Superfund project, for example, made use of sensors to control slurry density and velocity to meet the requirements of the processing facility (Taylor, 1995). Except in rare circumstances, temporary storage needs to be provided to accommodate fluctuations in sediment loading rates. Because many treatment operations also require pretreatment, it may be economical to combine pretreatment and temporary storage.

Precision and Accuracy of Removal

The requirements and objectives of routine maintenance dredging differ somewhat from those of environmental dredging. The differences in the precision and accuracy of sediment removal are striking. The cost of conventional dredging operations is usually based on the amount of sediment removed from a defined volume called the prism. Because the contractor bears the costs of dredging outside the pay prism, excessive dredging is seldom a problem, assuming the additional placement costs are minimal. But the removal of contaminated sediments very often requires special handling, treatment, and placement, and unit costs may be much higher than routine dredging costs. Therefore, overdredging of contaminated sediments should generally be avoided if at all possible.

It is important to realize, however, that precision sediment removal must be appropriate for the degree of site definition; conversely, the characterization of the site must match the precision of the available dredging equipment. Time and money are often wasted in the precise mapping of layers of contaminated sediment to the centimeter when the dredging equipment removes layers tens of centimeters thick. Conversely, there is no reason to remove sediments with a precision of a few centimeters if the contaminated layers have not been defined to a similar level of precision. Although precise removal may lower the cost of treatment, it also raises the cost of site assessment, principally because of the high cost of the chemical testing of sediment samples. If contaminated and uncontaminated sediments exhibit distinctively different physical properties, acoustic profiling systems (discussed in Chapter 4) may increase the precision of site assessment at a reasonable cost, thereby helping to reduce the volume of uncontaminated sediments removed. The degree of precision that is cost effective must be determined on a site-specific basis.

Most of the equipment in the U.S. dredging fleet is capable of removing sediments with a horizontal precision of a few inches in relatively shallow water under calm conditions. But horizontal precision may decline to around 2 ft in deeper water or heavy seas. The horizontal accuracy of the dredge cut depends largely on the system used to position the dredge. For the past decade or more, microwave positioning systems have provided horizontal accuracies of ±5 to 10 ft and have often been used to position dredges. Recently, the satellite-based global positioning system (GPS) has provided similar horizontal accuracies when the differential mode is used (U.S. Department of the Army and USACE, 1995). By 1995, GPS technology had advanced to the point that differential GPS could provide three-dimensional accuracies of ±0.3 ft or better (Frodge et al., 1994). Thus, differential GPS, when routinely applied, should allow horizontal positioning in the 0.5- to 1-ft range required for moderately to highly contaminated sediments when overlap between successive passes of the dredge is critical.

Laser-based positioning systems can easily provide horizontal positioning of less than 0.2 ft (Clausner et al., 1986). For example, the closed-bucket dredge used at Bayou Bonfouca used laser-based positioning systems to achieve accuracies of 0.1 ft (Thoma, 1994). But the accuracy of laser-based positioning systems degrades with distance from the reference station. The system was designed for use in relatively low-energy environments in water depths of less than 40 ft.

Vertical precision is generally greater than 0.5 ft for most conventional, fixed-arm dredging equipment. Conventional bucket dredges, however, operate by dropping the bucket into the bottom sediments, with minimal control over the depth of penetration. The actual depth of penetration depends on the strength of the bottom sediments. This problem can be overcome with some new bucket designs, such as the cable arm clamshell (see Recent Dredging Innovations below), which can leave a relatively smooth, flat bottom by monitoring vertical penetration with pressure sensors and depth sounders.

The most critical aspect of positioning for the removal of contaminated sediments is the vertical accuracy of the dredge-head. In many cases, contaminated sediments are concentrated in thin layers measuring in the tens of centimeters (van der Veen, 1995). To remove the thin layers of contaminated sediment with a minimum of additional clean sediments requires precise control of the elevation of the dredge-head relative to a fixed datum. For floating platforms, the reference datum is the water surface. Because the elevation of the water surface can vary, often considerably over a short period of time when the area is influenced by tides, accurate knowledge of the water surface is critical. Fortunately, most areas with significant contamination are in rivers, estuaries, and harbors, where surveyed ground elevations and accurate tide gauges can provide the necessary information.

Bottom-crawling platforms with hydraulic dredging systems, similar in concept to those developed for deep-ocean mining, can be positioned precisely. Because these platforms contact the bottom sediments directly, they are not subject to waves or traffic, which can complicate the positioning of surface platforms. Although bottom-crawling systems have not been used in the United States, they have been used in Europe. They are also readily available, moderately priced, and usable for dredging contaminated sediments. Some concern has been expressed that soft sediments may not be able to support bottom-crawling equipment; however, this problem can be overcome by replacing the standard track mobility system with Archimedes screws for traversing soft sediments and slopes (Wenzel, 1994a). Bottom-crawling systems are sensitive to the relief of the bottom topography. There are also concerns that the propulsion tracks could cause mixing of spilled contaminated sediments with clean sediments (van der Veen, 1995), but this problem can be overcome by adjusting the cutter width to greater than the track width.

In the future, the depth of cut might be automatically controlled with sensors, such as highly accurate acoustic sensors (discussed in Chapter 4) that can measure the thickness of contaminated sediments (Caulfield et al., 1995; McGee et al., 1995). The use of acoustic sensors would require that contaminated and uncontaminated sediments have significantly different physical properties; such differences were demonstrated in the Trenton Channel of the Detroit River (Caulfield et al., 1995). Acoustic sensors could also be used for dredge-heads deployed from surface platforms.

Characteristics of Delivered Material

In practice, uncertainties about the character of a site, the treatment approach, and the availability of appropriate dredging equipment make it difficult to find the right match between the requisite precision of the site assessment and the dredging technology. Therefore, the cost and magnitude of treatment must be calculated based on characteristics of the material as delivered, as opposed to characteristics in situ. The mechanical action of removing sediment and the imprecision of current dredging technology raise concerns about contaminated upper layers mixing with cleaner lower layers. Mixing dilutes the contamination in the dredged sediment and increases the volume that has to be dredged.

Volume reduction requires the removal of only those sediments requiring treatment and the entrainment of as little water as possible during the removal process. Mechanical dredging tends to keep water content low. But there are other alternatives. The dredge developed for the Bayou Bonfouca project (described above) included a patented processing unit to control the density of the hydraulically transported slurry within limits acceptable to the sediment treatment facility. Some foreign technologies are said to be capable of removing sediments at very high, even near-in-situ, densities, but definitive data are not readily available to validate these claims, probably because the information is proprietary. There is a significant need for the demonstration of operational hardware developed overseas as well as a need for more U.S. R&D focused on improving dredging precision and accuracy, methods of delivering undiluted contaminated sediments to treatment facilities, and the use of acoustic sensors for site characterization and dredging control.

Environmental Risks

The disturbances created by removing and transporting contaminated sediments may increase the risk of contaminant release to the environment. This section summarizes the results of recent research into the extent of these releases and discusses ongoing technology developments in this area.

Contaminant Release Associated with Dredging

Most contaminants associated with sediments tend to remain tightly bound to fine-grained particles and controlling their resuspension is a key consideration in controlling contaminant releases from dredging. The strong hydrophobic nature of most contaminants associated with sediments suggests that releases of dissolved contaminants into the water column are minimal (Digiano et al., 1993). Resuspensions of contaminants from dredging are generally local, and the acceptable level of resuspension is a site-specific issue. To determine whether a predicted level of resuspension is acceptable, the dredging operation must be viewed as part of a whole that includes existing site conditions (e.g., the level of resuspension from ambient currents, storms, flood flows, etc.), the location and nature of resources of concern, and potential releases from other pathways associated with the disposal alternatives under consideration (Palermo et al., 1993).

In the NRC report (1989), field and laboratory studies quantifying the extent and mechanisms of sediment resuspension were summarized. The studies show that resuspended sediment concentrations are generally less than 100 mg/L except in the immediate vicinity of the dredging operation. In most of the field studies, resuspended sediment concentrations were less than 10 mg/L at distances on the order of 100 m from the dredge. These results contrast sharply with some early assertions of resuspended sediment concentrations of more than 1,000 mg/L. Very high concentrations were observed in some laboratory studies (Herbich and DeVries, 1986), but they probably reflect scaling difficulties between hydraulic parameters and sediment settling rates. None of the field studies conducted in the United States has revealed such high suspended sediment concentrations (McLellan et al., 1989; Collins, 1995).

Field studies conducted to date provide valuable site-specific information, but the results are difficult to apply to other sites or to equipment being used under different conditions. Many of the studies involved monitoring resuspension generated by a single dredge type operating at a specific site. However, several of the studies involved comparisons of the sediment resuspension from two or more dredges of different designs operating at the same site (Hayes et al., 1988; Otis, 1992). A few attempts have been made to develop generalized predictive tools from field data so that a priori estimates of concentrations of resuspended sediment can be developed systematically (Bohlen, 1978; Cundy and Bohlen, 1982; Herbich and DeVries, 1986; Crockett, 1993; Collins, 1995). Collins (1995) developed mathematical models for predicting rates of sediment resuspension for conventional bucket and cutter-head dredges for a limited range of conditions at a few sites. Designing models is difficult because the vast number of operational choices for each dredging operation and the disparity in conditions among field studies mean that very sparse data are available for evaluating a large array of possible combinations. The models that have been developed are mostly unverified (Collins, 1995).

Based on the available data, it appears that the total amount of sediment "lost" to resuspension is 2 to 5 percent of the in situ volume. However, this small percentage does not necessarily mean that sediment resuspension is not a concern: 1 percent of certain contaminants could be a substantial problem. The presence of debris, ranging from household garbage to logs and automobiles, in the sediments can increase resuspension by interfering with the dredging process. The backhoe dredge used at Bayou Bonfouca was particularly well suited to working in sediments mixed with large wood fragments, which were separated out prior to sediment processing.

No unusual problems were associated with the dredging of highly contaminated sediments from the New Bedford Harbor Superfund site. Otis (1992) ran a pilot study using three types of dredges. Regarding the dredging of the PCB-contaminated hot spot, Otis (1994) reported "no problems with sediment resuspension or contaminant release in the water column" using an extremely slow production rate. This result was upheld by laboratory studies (Digiano et al., 1993, 1995) examining the partitioning of contaminants to estimate the potential release of PCBs during dredging operations and comparing the results of the pilot study. Monitoring during the New Bedford pilot study verified that releases of dissolved contaminants were rather small (Otis, 1992).

Specialty dredges have been designed to reduce resuspension during dredging operations, and many are effective in removing sediment with a minimum of resuspension. However, field tests indicate that conventional dredges, if operated with care, can also remove sediment with low levels of resuspension (Hayes et al., 1988; Otis, 1992).

Contaminant Losses during Transport

Some contaminants may be lost during certain phases of sediment transport. For purposes of the following discussion, the transport system includes all operations between sediment removal and delivery, up to the point of ex situ treatment or placement. Hydraulic-based delivery systems are essentially "closed" systems with no significant opportunities for contaminant losses, except at the point of discharge (assuming there are no leaks or breaks in the pipeline).

Mechanical dredging systems and some hydraulic systems use a hopper barge or vessel to deliver sediments for ex situ treatment or disposal. Once sediments are placed in the hopper, they settle to a considerable degree, resulting in a dense sediment load near the bottom and free water on the top. In conventional dredging operations, it is common practice to continue loading the hopper until much of the free water has been displaced by sediment. This practice is known as "increasing the economic load." The free water overflows the hopper and is discharged directly into the water column. But even with sediments containing low levels of contamination, the carryover of fine-grained material can be a problem because there is seldom time to permit settling. Overflow is a source of water-column turbidity and potential contaminant loss. Although the amount of water-column turbidity attributable to overflow has not been quantified directly, some researchers have estimated that the amount is comparable to the amount from the dredging operation itself (Hayes, 1993). Overflow is avoidable but requires more hopper loads. If sediment resuspension must be reduced, then hopper overflow can be minimized.

When sediments are transported by hoppers, they must be mobilized again at the point of ex situ treatment or placement, with some potential for contaminant loss. One mechanism for loss, vaporization of contaminants, is seldom a concern because very few volatile contaminants are likely to be associated with sediments. However, at the New Bedford site, volatilization of PCBs from the placement facility has been a concern during the dredging operations (Thibodeaux, 1989; Otis, 1994); and if this material had been transported by barge, then PCB emissions would have been an issue. Sediments with the potential to release free sulfides during transport and handling may be a nuisance if not a contamination concern.

On-Site Controls

Regardless of the control measures, some contaminated sediments will escape from the dredging operation. Fine sediment particles can be transported from the dredging area by even relatively slow currents, and fine particles have the highest affinity for hydrophobic contaminants. Concern about the environmental impact of in-place contaminated sediments is often exceeded by anxiety over the potential spread of contamination to down-current areas. A risk-based assessment may be one way to put these concerns in the proper perspective. The monitoring of dredging operations has shown that such concerns are usually exaggerated and that, in general, the amount of sediment transported off site is very small. Nonetheless, transport can be minimized by using good engineering practices.

The most common method of isolating a dredging area involves the use of silt curtains, but they require such special conditions for successful operation that they are rarely effective. Silt curtains are made of geosynthetic fabric and are hung vertically from a floating support. The fabric may be either impermeable or porous (often referred to as a silt screen). Provisions must be made to ensure that impermeable curtains permit currents or tidal flows to pass underneath them, around them, or through windows in the curtain. Silt screens are intended to filter sediment particles as the water passes through the openings, even though the pores in the fabric are typically much larger than the particles of concern. Silt screens can be effective when secured well enough to force water to flow through the small openings, but this is usually possible only in areas with very low currents and low winds or areas where the curtains can be fastened securely to bulkheads or piers. Even at low flow rates, water will pass underneath the curtain unless the fabric is anchored securely; in modest currents, it is almost impossible to anchor the curtains sufficiently.

An alternative, three-step approach that also has very limited application involves the physical isolation of the dredging area by using sheet piles or cofferdams. Dredging can be performed inside the cofferdam, or the area can be dewatered and the dry sediment excavated. A sheet-pile wall was used at a cleanup site on the Saint Lawrence River to isolate a dredging area along the shore. The sheet-pile wall was used because the current precluded effective deployment of silt curtains.

INTERIM AND LONG-TERM TECHNOLOGIES AND CONTROLS

Dry excavation is a precise but expensive method that is warranted only when small hot spots have to be removed for extensive treatment. In these instances, the high set-up and removal costs might be partly offset by reductions in dredging and treatment costs. At a Superfund site in Cedarburg, Wisconsin, the flow in Cedar Creek was diverted into pipelines, and a 1,000-ft segment of the riverbed was drained so that 25,000 yd^3 of PCB-contaminated sediments could be excavated with conventional earth-moving equipment (J. Miller, USACE, personal communication to Marine Board staff, June 7, 1996). The advantages of this method are twofold. First, removal equipment can be operated with great precision when the operator actually can see the sediment being removed. Second, after dewatering sediment can be removed with far less water entrained than with routine dredging. A disadvantage is the increased potential for contaminant volatilization because the sediment is exposed to the air.

A pneumatic barrier, consisting of bubbles from a submerged pipe, has been used to contain oil spills and has been proposed for use in managing contaminated sediments in Boston Harbor. The system can be deployed and maintained easily, and the absence of near-surface physical structures, such as floating booms, permits the free passage of vessels; the pneumatic barrier may also cost less than silt curtains. However, an air barrier was used with poor results during dredging at Indiana Harbor in the late 1960s (J. Miller, USACE, personal communication to Marine Board staff, June 7, 1996). It is not clear how aeration affects contaminant release from resuspended material.

Recent Dredging Innovations

Fundamental dredging equipment and methods for efficiently moving large quantities of sediments have not changed substantially in several decades. However, a number of equipment enhancements and specialized dredges have been developed specifically for dredging contaminated sediments (NRC, 1989; Herbich and Brahme, 1991; Zappi and Hayes, 1991; EPA, 1994b). This section updates the 1989 NRC report. Detailed comparisons of various dredge types are available elsewhere (Herbich, 1995; van der Veen, 1995).

Many of the technologies described in the 1989 NRC report were of Dutch and Japanese origin. Despite the prohibition on using foreign-flag dredges in U.S. waters, (Jones Act, 46 CFR §292), the committee does not view this as a major problem because most foreign innovations are in the dredge-head rather than the platform. But access to foreign technologies is not the primary barrier to improved sediment handling because, although Dutch and Japanese development has continued, many of the advances described in this section were developed in the United States. The momentum probably shifted to North America simply because of the demand for the equipment created by several contaminated sediment dredging projects, notably around the Great Lakes (both the U.S. and

Canadian sides).[6] It is important to note that, although numerous articles have reported field observations related to new equipment (Otis, 1992; Buchberger, 1993; Kenna et al., 1994; Pelletier, 1995), dredging projects seldom monitor or document operating conditions or sediment characteristics, perhaps because of the expense involved. Thus, a number of commercial dredging innovations must be considered unproven until additional data become available.

One recent innovation is the cable arm environmental clamshell, which is used for sediment removal by bucket dredges. The distinguishing feature of the cable arm is its capability of removing sediment and leaving a horizontal bottom prism rather the cratered prism left by most dredge buckets, which tend to overdredge. The cable arm was modified recently with a vertical side plate to prevent the lateral flow of contaminated material from the bucket during the environmental dredging of contaminated sediments. The cable arm has been used for a number of projects in the United States and Canada. Buchberger (1992, 1993) described the use of the cable arm clamshell in Toronto's Inner Harbor. Water quality studies conducted during the Toronto study did not indicate any unusual environmental problems from the use of the cable arm clamshell compared with traditional clamshell buckets (Buchberger, 1993).

Concern about the precision of sediment removal and dredge-head positioning in water much deeper than is usually encountered in navigation dredging led Wenzel (1994b) to recommend the use of a bottom-crawling dredge for the cleanup of contaminated sediments on the Palos Verdes shelf and slope. The contaminated sediments were spread in a thin layer (30 to 60 centimeters [cm]) over a large area (approximately 16 square kilometers [km^2]) in waters 30 to 500 m deep. Cost considerations required precise vertical dredging control, raising concerns that surface dredges would have difficulty accurately removing the sediments from such deep water. A bottom-crawling dredging system was selected that had been used to clear contaminated sediments from around oil field platforms in the North Sea (Alluvial Mining Group, Ltd., 1993) and had been effective in placer mining in the shallow waters of Alaska. Although this technology has not been demonstrated at contaminated sediment sites in the United States, there are no known impediments to using it in this context. To determine whether it offers benefits that justify additional costs, a side-by-side comparison with dredges currently available in the United States would be useful (M. Palermo, USACE, personal communication to Marine Board staff, December 15, 1995).

In 1992, van Oostrum described a conceptual approach to dredging in which sensors were used to determine the horizontal and vertical location of contaminated sediments and to guide the dredge to remove only those sediments. This approach was termed "digital dredging." Although the approach remains conceptual, recent

---

[6] Environment Canada has worked extensively with dredging contractors to develop and demonstrate innovative equipment as part of the Contaminated Sediment Removal Program of the Great Lakes Cleanup Fund.

success in sensor development (see Chapter 4) suggests that digital dredging may eventually become an implementable system. A digital system could control the quantity of uncontaminated sediments removed and thereby reduce overall remediation costs (van Oostrum, 1992). Sensors have also been proposed for use with bottom-crawling systems to ensure, and provide legal verification of, the removal of contaminated layers in a single dredging pass (Wenzel, 1994b).

Van Oostrum (1992), Kato (1993), and Keillor (1993) discuss the importance of limiting water entrainment during the dredging process. Some foreign hydraulic dredging technologies have been touted as being capable of removing and transporting sediments at near-in-situ density. However, none of these technologies has been proved thus far to accomplish this under normal circumstances (McLellan et al., 1989). Kato (1993) describes two dredge models based on a conveyor design that attempts to increase solids content in the delivered slurry. Although laboratory tests of these models seem promising, the designs have not yet been tested in larger-scale units.

Another fairly recent innovation is diver-assisted dredging, which is being used at two Great Lakes sites to minimize the resuspension of contaminated sediments. Divers holding small-diameter pipelines connected to a suction pump are removing approximately 100,000 yd^3 of sediments from a water intake flume in Indiana Harbor (J. Miller, USACE, personal communication to Marine Board staff, June 7, 1996). At the Manistique Harbor Superfund site, diver-assisted dredging is being used to remove approximately 1,900 yd^3 of PCB-contaminated sediments.

A useful summary of equipment developed (as of 1991) for dredging contaminated sediments can be found in a report commissioned by the Directorate-General for Public Works (Rijkswaterstaat) of the Dutch government (van der Veen, 1995). The report examined primarily Dutch dredges but also included some equipment developed in the United States and Japan. More than 40 dredges were rated with respect to the concentration and density of the supplied slurry, the accuracy of vertical selectivity, the accuracy of horizontal overlap, dispersal (turbidity generated by the dredge), the mixing of contaminated material with subsoil, the clearing of spillage, and crew safety. The overall conclusion was that a number of dredges can be used for the environmentally effective removal of contaminated dredged material. The best systems combined mechanical methods to loosen the sediments with hydraulics to transport the sediment to the surface. These dredges include auger dredges with screens, disc cutters with screens, shoveling suction silt plows, conventional auger dredges, and cutter suction dredges with Otter heads.

The report found that specialized equipment is required for the removal of thin layers (0.5 m or less), whereas carefully operated conventional equipment can be used for thicker layers, if resuspension is not a problem. The report recommended the continued development of systems that can deliver highly concentrated sediments and that incorporate advanced process control by environmentally aware operators.

Industry experts suggest that continued advances in equipment or operational approaches would be promoted if innovation were encouraged in the contractor selection process for contaminated sediment removal (A. Taylor, Bean Dredging, personal communication to Marine Board staff, December 12, 1995). Some recent dredging contracts have specified the equipment and approach in an effort to streamline the permitting and approval process by local or state environmental agencies. This practice is perceived to discourage private-sector innovation. Both the contractors and their clients might be better served if the site-specific problems were well defined and procurement was based on a performance specification allowing the contractor to investigate, develop, and offer scientifically proven solutions based on experience and testing.

The dredge developed for use at Bayou Bonfouca is an example of a dredge system tailored to project specifications, which, in this case, required sediment removal from the defined prism with a tolerance of only –0.5 ft. The dredge was outfitted with sensors that allowed bucket positioning and sediment removal with an even higher level of precision than was required. The environmental characteristics of the dredge have not been evaluated independently, but monitoring during the sediment removal process and follow-up surveys indicate that the project was accomplished within the criteria set forth by the EPA (Taylor, 1995).

TECHNOLOGIES FOR EX SITU MANAGEMENT

If dredged and transported sediment is too contaminated for open-water disposal, it may require treatment or containment. Treatment processes attempt to physically, chemically, thermally, or biologically alter contaminants through concentration, isolation, destruction, degradation, or transformation. Containment systems are designed to remove the residuals from contact with the biologically accessible environment and to minimize contaminant losses from their boundaries. The following practical appraisal of general approaches is intended to serve as a guide in the evaluation of management options (See Box 5-3, Selecting Ex Situ Controls.). Approximate cost data are provided where available.

In a systems approach to remediation, ex situ management costs must be added to the costs of dredging, transport, and disposal. Costs can be increased further by the need for interim storage facilities. This need is driven by two factors. First, the optimal processing rates of dredging, treatment, and disposal technologies may not be compatible. To be economical, dredging operations are done at a high rate and nearly continuously. The slurry is usually produced at a flow rate and with a water content that are not suitable for immediate input into a treatment process. Thus, interim storage facilities are needed to accommodate the production rates of the treatment facility, which are usually much slower than the dredging rates. Treatment processes are best operated on a steady-state basis with nearly uniform feed characteristics, but the process(es) must be flexible so that

> **BOX 5-3**
> **Selecting Ex Situ Controls**
>
> Many ex situ technologies have been investigated, and rankings are available (e.g., Averett and Francingues, 1994).
>
> The key issues are cost and economy of scale; the challenge is to select the technology appropriate to the job at hand.
>
> Because dredged materials often contain multiple contaminants, a combination of treatments may be required, which will add to the cost.
>
> All treatment technologies involve complex chemistry, so case-by-case treatability studies are required.
>
> Management plans that incorporate treatment technologies need to account for the proper disposition of all waste streams, including aqueous and gaseous releases, cleaned solids, solvents, and concentrated residuals.

changes can be made in response to operational problems. The second driving factor for using interim storage is the frequent need that different treatment processes be carried out sequentially. (Additional information on treatment technologies applicable to contaminated sediments can be found elsewhere [Averett et al., 1990; EPA, 1993a, 1994b; Tetra Tech and Averett, 1994].)

Treatment Systems

Although considerable research has been done on the treatment of contaminants, particularly in soil, the ex situ treatment of contaminated sediments is still very expensive and has been used only at a dozen or so sites in North America (Averett and Francingues, 1994).

In the design of treatment systems for complex wastes, particularly when large sediment volumes are involved, the standard approach is to perform the simpler, easier, and less-expensive processes (e.g., particle size separation) first and the more difficult or more energy-intensive processes later. Because organic contaminants tend to associate with fine-grained sediments, particle separation by size could be carried out first to reduce the volume to be treated, provided the grain size distribution and contaminant distribution favor separation. Treatment processes requiring changes in temperature or additions of reagents work most efficiently on low volumes of highly concentrated materials.

After any sediment treatment process, placement sites must be found for large volumes of sediment and water. Small sediment volumes with highly concentrated contaminants can be isolated or destroyed using expensive processes, such as landfilling or incineration. Cleaner sediment can be put to beneficial use and may even have a market value (as discussed in Chapter 3), or can be placed in open water.

Residues remaining after treatment must be evaluated against regulatory standards to determine suitable placement alternatives, which may include hazardous waste landfills. Landfill costs vary, ranging from $20 to $24/yd^3 for nonhazardous solid waste to $120/yd^3 for waste classified as hazardous (EPA, 1994b). The USACE is investigating whether treatment residues can be put to beneficial uses as components of soil, bricks, or road aggregates (C.R. Lee, U.S. Army Engineer, Waterways Experiment Station, personal communication to Marine Board staff, December 18, 1995).

Five general ex situ treatment processes are described below: pretreatment and solids-water separation; physical separation; chemical separation, thermal desorption, and immobilization; thermal and chemical destruction; and biological treatment.

Pretreatment and Solids-Water Separation

The separation of solids from water is the simplest treatment process. The solids content of sediments varies with the technology used to recover them. Hydraulic dredges remove sediments in a liquid slurry that usually requires dewatering. Mechanical and pneumatic dredges remove sediment with solids contents at or near in situ levels. The dewatering of dredged material typically is accomplished in ponds or CDFs, which rely on seepage, drainage, consolidation, and evaporation (USACE, 1987). Dewatering is generally effective and economical, but slow, and the water generated, which usually contains contaminants, may also require treatment. Common industrial methods of dewatering slurries or sludges include centrifugation, filtration and filter presses, and gravity thickening. But these approaches are of limited value for sediments that contain silt- and clay-sized particles (EPA, 1993b).

Physical Separation

Soil washing and particle separation techniques are adaptations of mineral processing techniques used in the mining industry (see Galloway and Snitz, 1994). Soil washing is a general term for extraction processes that use a water-based fluid as a solvent; many soil washing processes rely on particle separation (EPA, 1994b and references therein). The state of the art is summarized in Table 5-6.

Particle classification separates sediment particles based on one or more physical properties, such as size, density, or surface chemistry. In both freshwater

TABLE 5-6 Soil Washing and Physical Separation

State of Practice (system maturity, known pilot studies, etc.)	Applicability	Advantages/Effectiveness	Limitations	Research Needs
Well developed by mining industry and frequently used for sediments.	Where contaminant is predominantly associated with fine-grained material that is a small fraction of the total solids.	(a) Mature technology that can reduce volumes of contaminated material requiring subsequent treatment; (b) soil washing can be used to recover CDF space for later reuse.	Original sediments must have a significant proportion of sand for the process to be cost effective.	None identified.

and marine sediments, contaminants are associated mainly with the silt- and clay-sized fractions rather than with sandy material (Gibbs, 1973; Moore et al., 1989). For example, in samples of sediment from the Saginaw River, 80 percent of the PCBs were associated with the finest-grained 20 percent of the sediment (Allen, in press). Sand separation from silt and clay-sized material is achieved with hydrocyclones, in which particles exposed to a centrifugal field settle at size-dependent rates. In principle, particles larger than 0.0062 mm in size can be separated from dredged sediments by screening, but in practice separations are easier for particles larger than approximately 1 mm in diameter. At Manistique Harbor, screens were used to separate dredged sediments from wood chips, which contained a high concentration of PCBs. Sometimes schemes are combined. An example of multistage physical separation is the process used at the largest particle separation system for dredged material in the world at the Port of Hamburg in Germany, where all dredged sediments from the highly contaminated Elbe River are pretreated. The system uses screens, hydrocyclones, and belt filters to separate sand from silts and clays (Detzner, 1993).

Soil washing techniques can be used to recover storage space, which can be a useful sediment management strategy (see section on interim controls, above). This approach was demonstrated at a CDF in Michigan, where sediments from the Sagninaw River contaminated with PCBs and metals were separated into a large volume of fairly clean sand and a small volume of fine sediments containing the bulk of the contaminants (U.S. Army Engineer Detroit District, 1994). Soil washing has been used routinely at a CDF in Duluth, Minnesota, to reduce the volume of dredged sediments requiring confined disposal (Miller, 1995). Soil washing results in a large volume of "clean" material, which can be put to use, and a small, concentrated amount of highly contaminated material, which must be disposed of. Clean, sandy sediment can have a wide range of uses in urbanized coastal environments and may be more readily available than other sources of sand. Unfortunately, the sand fraction for most contaminated marine sediments is a small percentage of the total.

Physical separation can be facilitated by differences in surface chemistry. The minerals processing industry routinely separates desirable minerals from crushed rocks by adsorbing surfactants on the minerals of interest and selectively recovering the ore by flotation. Surfactants also have been used to solubilize more than 95 percent of the oil from contaminated sediments and to remove a comparable percentage of PCBs because the PCBs were strongly partitioned within the oil phase (Allen, in press).

The cost of physical separation depends on the number of steps and the volume of sediment. For a sediment containing 75 percent clean sand and 25 percent contaminated silts and clays, the costs of physical separation using a system of screens, trommels, hydrocyclones, attribution scrubbers, and other equipment are estimated at \$23 to \$54/yd^3 for a volume of 10,000 to 100,000 yd^3 (U.S. Army Engineer Detroit District, 1994). In general, physical separation is worth the expense only if the contaminated sediment is at least 25 percent sand[7] (D. Averett, USACE, personal communication to Marine Board staff, January 2, 1996).

It is important to emphasize that separation is not an effective treatment for all sediments and does not destroy the contaminants but concentrates them into a smaller volume, leaving a large volume of only slightly contaminated sediment. The reduced volume of concentrated waste may be suitable for high-energy chemical, thermal, or biological treatment, if the benefits outweigh the costs. Reductions in volume also lower handling and disposal costs.

Chemical Separation, Thermal Desorption, and Immobilization

The contaminants accumulating in bottom sediments preferentially associate with fine particles rather than dissolving in the water. Chemical separation and thermal desorption processes attempt to mobilize these contaminants into a fluid or gas phase where the contaminants can be concentrated, isolated, or destroyed. Key considerations of these processes are summarized in Table 5-7.

For the removal of metals, the fluid phase can be a leaching solution composed of an acid, a base, or a metal chelator. Acid leaching is a convenient method of dissolving basic metal salts, such as hydroxides, oxides, and carbonates. (Basic solutions are also usable for releasing certain metals adsorbed to mineral surfaces.) The leached sediments require neutralization following metal dissolution, and the aqueous solution typically must be clarified to remove the suspended particles. The extracted metal-rich solution can then be concentrated by precipitation or ion exchange. Overall cost estimates for metal leaching are on the order of $120 to $200 per ton (EPA, 1993b). The efficiency of this process during bench-scale testing was, at most, 75 percent from a sandy sediment (Wardlaw, 1994). Leaching sediments containing metals present as sulfide precipitates would be ineffective, however, given the low solubility of the precipitates.

The separation of organic contaminants requires a nonpolar phase, such as hexane, chlorofluorocarbon, triethylamine, or supercritical carbon dioxide and propane. The extraction liquid must be mixed vigorously with the sediment to achieve equilibrium, and then the liquid and sediment are separated. Repeated washings are needed to remove contaminants efficiently. These sediment washing processes are done in batch reactors, with 50 to 75 percent contaminant removal during each cycle (efficiency rates are limited by fluid carryover during solid-fluid separation). To achieve 99 percent contaminant removal, four or more sequential washes are required. The process is cost effective only if the contaminants can be separated from the extracting liquid, and the extracting liquid can be reused. Particularly useful in this regard are supercritical fluids, which can be

---

[7] In most cases, the sand fraction is low (3 percent or less). However, some locations in the United States have a volume of sand sufficient to yield a fraction of greater than 10 percent.

TABLE 5-7 Chemical Separation and Thermal Desorption

State of Practice (system maturity, known pilot studies, etc.)	Applicability	Advantages/Effectiveness	Limitations	Research Needs
(a) Pilot plant studies conducted on metal desorption by acid-leaching solutions and at least one full-scale implementation; (b) pilot and full-scale application of organics separation by liquid solvents and supercritical fluids; (c) organic chemical thermal desorption also has had full-scale demonstration; (d) thermal desorption used at Waukegan Harbor.	Suitable for weakly bound organics and metals.	Contaminant is removed and concentrated.	(a) Batch extraction during separation requires multiple cycles to achieve high removal; (b) fluid-solid separation is difficult for fine-grained materials; (c) a separate reactor is needed to remove the contaminant from the extracting fluid so that the extracting fluid can be reused; (d) thermal desorption requires temperatures that will vaporize water, and sediment particles must be eliminated from gaseous discharge; (e) contaminant removal from the gas phase following thermal desorption is another treatment process that is required.	Systems integration for complete contaminant isolation or destruction.

used to extract organic contaminants at high pressures and can then be separated out easily for reuse later by restoration of atmospheric pressure, at which the supercritical compound is a gas. This approach was used at pilot scale in New Bedford (EPA, 1990, 1994b). Costs are estimated at $140 to $360/yd^3 (EPA, 1994b).

Volatile and semivolatile organic contaminants also can be vaporized from sediments at temperatures of 200°C to 300°C using any of a number of proprietary thermal desorption technologies. The resulting off-gases are first treated to eliminate the dust and then cooled. This is followed by condensation, which produces water and an organic vapor phase. The two liquid streams, in addition to the remaining gas stream, require further treatment and disposal. Energy costs depend on initial moisture content, which must be less than 70 percent to ensure cost effectiveness. Pilot testing with this technology revealed problems with materials handling, and costs were estimated at $270 to $540/yd^3 (U.S. Army Engineer Buffalo District, 1993, 1994). At the Waukegan Harbor Superfund site, thermal desorption was used to reduce PCB concentrations in excess of 500 ppm to less than 2 ppm in the residual sediments at a cost of approximately $250/yd^3, plus fixed costs of $150/yd^3 (see Appendix C).

An alternative process is chemical immobilization, which involves chemically isolating contaminants from the biologically accessible environment. The state of the art is summarized in Table 5-8. Chemical immobilization by solidification converts sediments into solid blocks by the addition of cement, silicates, and proprietary reagents. Some stabilization processes adsorb or react with free water in the sediment to form a relatively dry material without hardening into a monolith. The stabilization process used at the Marathon Battery Superfund site produced a soil-like material that reportedly immobilized the metals in the sediment. Water contents below 50 percent are probably desirable to make the process cost effective. Solidified volumes can be up to 30 percent larger than the initial sediment volume. Estimated costs for solidification and immobilization range from $50 to $150/yd^3 for total containment of metals.

Whether this approach is effective for treatment of organics is unclear (Averett et al., 1990; EPA, 1993a). Laboratory experiments with New Bedford sediments showed that solidification successfully reduced the mobility of metals (Myers and Zappi, 1992). This approach has several benefits, including simplicity, a history of use with sludge, and the capability of improving handling of sediments. However, the solidified material must still be disposed of.

Thermal and Chemical Destruction

Heat or chemical reactions can be used to break down organic molecules into less hazardous forms. Thermal destruction is the most widely used destruction technology for organics and has achieved very high removal efficiencies—but at high costs. These and other considerations are summarized in Table 5-9.

TABLE 5-8 Immobilization

State of Practice (system maturity, known pilot studies, etc.)	Applicability	Advantages/Effectiveness	Limitations	Research Needs
Extensive knowledge based on inorganic immobilization within solid wastes and dry soils.	Chemical fixation and immobilization of trace metals.	(a) Chemical isolation from biologically accessible environment; (b) process is simple and there is a history of use for sludge.	(a) Sediment should have moisture content of less than 50 percent, and solidified volumes can be 30 percent greater than starting material; (b) limited applicability to organic contaminants; (c) high organic contaminant levels may interfere with treatment for metals immobilization; (d) need for placement of solidified sediments.	(a) Studies of long-term effectiveness for contaminant isolation; (b) develop sediment placement options, especially for beneficial uses.

Sediment incineration requires temperatures in excess of 900°C with gaseous contact times of a few seconds and solids with a contact time of minutes to hours, depending on the specific configuration. Treatment technologies are based on combustion in an oxidizing environment or reduction in a nonflame reactor. In combustion systems, the sediments are in contact with an oxidizing flame, and organic materials are oxidized to carbon dioxide and water vapor; if chlorinated materials are present, hydrogen chloride is produced as well. Fuel must be added for the incineration of sediments, given their low energy content, even if they have been dewatered. Post-combustion treatment systems include a secondary combustion chamber, gas quenching, particle-gas separation, and gas scrubbing for acid removal. These processes are followed by gas discharge, scrubber effluent treatment, and particle concentration and disposal.

In nonflame systems, such as pyrolysis and reductive dechlorination, heat is applied to the waste so that temperatures of 1000°C are approached to decompose the organic pollutants to carbon, carbon monoxide, hydrogen, dehalogenated organics, and hydrogen chloride. Following particle–gas separation, the gases undergo further treatment prior to venting to the atmosphere.

Thermal destruction technologies can achieve destruction and removal efficiencies of 99.99 percent for polyaromatic hydrocarbons and PCBs, but at costs ranging from $500 to $1,350/yd^3, depending on volume (EPA, 1993a,b). These removal efficiencies are upper limits because they are based on an analysis of stack gases only, rather than all residuals. When sediments are contaminated by metals, metal volatilization in combustion reactors results in metal condensation on fine particles. In these cases, further treatment is needed, and particle disposal options are more limited.

A number of chemical destruction technologies are under development for organic contaminants dissolved in water at ambient or elevated temperatures. Advanced oxidation processes based on ultraviolet light, ozone, hydrogen peroxide, and ultrasonics have achieved some success in treating halogenated organics present in water, but not on solid surfaces (Hoigné, 1988; Sedlak and Andren, 1994; Hua et al., 1995). Application of these technologies to the treatment of contaminated sediments presents many challenges related to ultraviolet penetration into slurries, oxidant demand by natural organic matter, the influence of metals and sulfides, process sequencing, and residuals management.

In experiments, PCBs have been destroyed by the nucleophilic substitution of chlorine by polyethylene glycol. The reaction is carried out at temperatures of 120°C to 180°C. A water content of less than 7 percent is required, along with a nitrogen atmosphere to keep the reagents from oxidizing. Residence time in the reactor ranges from 30 minutes to 2 hours, depending on contaminant characteristics and desired destruction efficiency. Problems that must be addressed prior to the large-scale application to dredged material include mixing of reagents, solids separation, reagent recovery and disposal, solids disposal, and treatment of the

TABLE 5-9 Thermal and Chemical Destruction

State of Practice (system maturity, known pilot studies, etc.)	Applicability	Advantages/Effectiveness	Limitations	Research Needs
Thermal oxidation in flame and thermal reduction in nonflame reactors have been extensively tested and demonstrated.	Process destroys organic contaminants in sediment samples at efficiencies of greater than 99.99 percent but at very high costs.	Very effective.	(a) Very expensive; (b) metals mobilized into the gas phase require gas phase scrubbing; (c) water content of sediment increases energy costs.	(a) process control to prevent upsets and effluent gas treatment for metals containment; (b) facility design to control the destruction process.

products of the organic reaction. Costs for nucleophilic substitution range from $200 to 500/yd^3 (EPA, 1993a,b).

Biological Treatment

Biological processes can be used in a number of ways to destroy or immobilize contaminants in dredged material. The simplest approach is land farming, where sediments are partially dewatered and occasionally tilled on the land surface to promote aerobic degradation. This approach is low in cost and has been used widely for treating soils contaminated by petroleum hydrocarbons. However, the process can take weeks or months and is not suitable for other contaminants, such as metals. Also available are more complex reactors containing slurried growth systems. Costs for all ex situ biological treatments are likely to be higher than costs for in situ alternatives because the sediments and other materials must be handled and greater energy is required for mixing. Ex situ treatment is also complicated by a number of other issues: large volumes of sediment must usually be treated; the sediments usually contain mixtures of organic and inorganic pollutants; the contaminant concentration is often relatively low; and aged polyaromatic hydrocarbons and PCBs are often less bioavailable than more recently sorbed compounds. Table 5-10 summarizes the relevant issues.

Some information on ex situ biological treatment is available from studies conducted on Zeebrugge Harbor, Belgium, at the bench, pilot, and demonstration scales (Thoma, 1994). The overall approach involved organic acid leaching for metals removal, followed by microbiological treatment for the degradation of polyaromatic hydrocarbons. Biotreatment consisted of land farming with the addition of nutrients, oxygen, surfactants, and degradative microorganisms. The result was a one-month half-life for contaminants and a suggested treatment time of six months at summer temperatures. Limitations of this approach, according to the researchers, include the need for site-specific feasibility studies and the limited volumes of sediments that can be handled.

Bioslurry reactors are a relatively new technology that has been used to treat contaminated solids (EPA, 1994b). There have been a number of pilot-scale applications in freshwater systems but few full-scale installations or demonstrations with marine sediments. For example, the degradation of PCBs using bioslurry reactor technology has been investigated for Hudson River sediments (Abramowicz et al., 1992) and tested in pilot-scale reactors for polyaromatic hydrocarbons (Toronto Harbor Commission, 1993). The results suggest that oil and grease are degraded within several weeks, with partial degradation of polyaromatic hydrocarbons.

At a Sheboygan River Superfund site contaminated with PCBs, ex situ bioremediation was demonstrated (on a pilot scale) in a CDF, which was constructed using large sheet-pile containment structures. A CDF is an ideal treatment facility for the bioremediation of sediments because it can be engineered to have controlled

TABLE 5-10 Ex Situ Bioremediation

State of Practice (system maturity, known pilot studies, etc.)	Applicability	Advantages/Effectiveness	Limitations	Research Needs
(a) Limited experience; (b) transfer of soil-based technologies to marine sediments is not proved and may not be directly applicable because of the different biogeochemistry of marine sediments; (c) but general trends should translate; (d) examples from freshwater sediment have been carried out at the pilot scale in the assessment and remediation of contaminated sediments program, as well as in Europe; (e) PCBs were treated ex situ at the Sheboygan River site.	(a) Contaminant is biologically available; (b) concentration of contaminant appropriate for bioactivity (e.g., sufficiently high to serve as substrate, not high enough to be toxic); (c) limited number or classes of contaminants are biodegradable; less known for complex mixtures; (d) site is reasonably accessible for management and monitoring; (e) rapid solution is not required.	Based on experience from freshwater systems, it offers the potential for (a) degradation (as opposed to mass transfer) of some organic contaminants; (b) possible reduction of toxicity from biotransformation in those cases in which complete mineralization does not occur; (c) containment of contaminated material allowing for an engineered system and enhanced rates, when compared to in situ biotransformations; (d) public acceptability.	(a) Far from a proven technology—all work with marine sediments is at the bench-scale; (b) requires handling of contaminated sediments; (c) slow compared to chemical treatment; (d) ineffective for low levels of contamination, and does not remove 100 percent of contaminants; (e) not applicable for very complex organics, such as high-molecular-weight compounds; (f) susceptible to matrix effects on bioavailability.	(a) Fundamental understanding of biodegradation principles in engineered systems; (b) exploration of aerobic/anaerobic combinations or comparisons; (c) laboratory, pilot, and field demonstrations; (d) analysis of cost effectiveness; (e) exploration of bioremediation as part of more extensive treatment trains.

conditions. The Sheboygan CDF was operated alternately as an anaerobic and then an aerobic digester to exploit the two-stage destruction of PCBs (EPA, 1994a) (outlined in the section on in situ bioremediation). The demonstration confirmed that the PCBs had undergone substantial anaerobic dechlorination before active treatment. Questions remain, however, about how to engineer a system that will deliver adequate amounts of oxygen to the sediments to break down the remaining, partially dechlorinated PCB molecules. Realistic cost estimates for this type of bioremediation cannot be made until the remaining questions concerning the design of a full-scale system have been answered (EPA, 1994a,b).

Treatment by composting[8] has been somewhat successful in a pilot project by Environment Canada's Clean Up Fund at a freshwater site in Burlington, Ontario. Approximately 150 tons of polyaromatic hydrocarbon-contaminated sediment from Hamilton Harbor were placed in a temporary shelter and tilled periodically with additions of organic matter (EPA, 1994b, and references therein). After an 11-month period, polyaromatic hydrocarbons were reduced by more than 90 percent in amended tillage, whereas controls (tilled but not amended) showed reductions of only 51 percent (EPA, 1994b, and references therein). However, controls with no tillage or amendment showed reductions of 73 percent. Research is needed to determine the mechanisms that led to these results.

Ex situ bioremediation, although not well developed, is considered to be more manageable than in situ bioremediation because it can be carried out in a contained environment, which, like a bioreactor, can be engineered to maintain controlled conditions. Indeed, ex situ bioremediation has many potential applications for the cleanup of contaminated environments and the treatment of hazardous wastes. It is generally recognized, however, that long-term programs and unusual efforts would be required to resolve the relevant R&D issues before treatment would be cost effective for contaminated sediments.

Effective bioremediation can reduce hydrocarbon concentrations in soil to levels that no longer pose an unacceptable risk to the environment or human health (Nakles and Linz, in press). Nevertheless, hydrocarbons that remain in treated sediment might not meet stringent regulatory levels, even if they represented site-specific, environmentally acceptable end-points. The availability of the remaining hydrocarbons is an unresolved issue that may influence the environmental acceptability of treated marine sediments. The development of standardized methods for assessing the availability for specific combinations of exposure routes and receptors will require joint efforts of the science, engineering,

---

[8] Composting is a biological treatment process in which bulking agents, such as wood chips, bark sawdust, and straw, are added to the sediment to absorb moisture, increase porosity, and provide a source of degradable carbon. Water, oxygen, and nutrients are added to facilitate bacterial activity. For sediments, dewatering may be a necessary pretreatment.

and regulatory communities because of the complexity of environmental systems and the interdisciplinary nature of bioremediation research.

Perhaps the most fundamental long-term issue to be confronted in bioremediation is the lack of understanding of contaminant-sediment interactions and their effect on the toxicity of contaminated sediments. Little is known about the mechanisms of chemical sequestration and contaminant aging in sediments and the resulting effects on chemical and biological availability. Long-term field studies of contaminated sites, with and without active bioremediation, are needed to evaluate the reductions in contaminant concentrations over time and to correlate these with reductions in availability, mobility, and toxicity. It is still not clear if reduced availability, biodegradability, and extractability correlate with reduced toxicity. Methods and protocols for measuring contaminant availability need to be designed in concept and then developed, validated, and standardized. The establishment of dedicated, well characterized field test sites and the establishment of postremediation monitoring requirements are subjects of ongoing debate.

Containment

Containment is a common approach to the ex situ management of contaminated sediments that have been dredged and transported. Ex situ containment has been widely used, at perhaps several hundred sites in North America. Containment technologies can be implemented in various ways. Figure 5-3 is an illustration depicting containment technologies, in situ capping, and deep-ocean dumping. The illustration highlights the distinctions among the different types of containment structures, particularly in terms of transport and isolating barriers. The subsections that follow assess CDFs, contained aquatic disposal (CAD), and landfills.

Confined Disposal

Confined disposal involves the placement of dredged material within diked near-shore, island, or land-based CDFs. Confinement or retention dikes or structures in a CDF enclose the disposal area above any adjacent water surface, isolating the dredged material from adjacent waters during placement. The enclosed disposal area of CDFs distinguish this disposal method from other disposal methods, such as disposal on unconfined land or placement on wetland or CAD, which is a form of subaqueous capping (USACE and EPA, 1992). The placement of dredged material in CDFs differs from the placement of waste materials in licensed solid-waste landfills (addressed in a forthcoming section).

The two objectives in the design and operation of CDFs that are used for contaminated sediments are to provide adequate storage capacity to meet dredging requirements and to maximize efficiency in controlling contaminant releases.

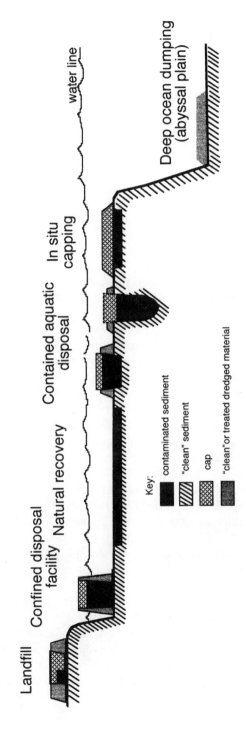

FIGURE 5-3 Conceptual illustration of containment, disposal, and natural recovery technologies. Dumping contaminated sediments in waters anywhere but in the open ocean is not permitted under the Marine Protection, Research and Sanctuaries Act.

Possible migration pathways of contaminants from CDFs include effluent discharges to surface water during filling operations and subsequent settling and dewatering, rainfall-generated runoff, leaching into groundwater, volatilization to the atmosphere, and direct uptake. Direct uptake includes plant uptake, subsequent cycling through food chains, and direct uptake by animals. Effects on surface water quality, groundwater quality, air quality, plants, and animals depend on the characteristics of the dredged material, the management and operation of the site during and after filling, and the proximity of the CDF to potential receptors of the contaminants. If evaluations of contaminant pathways indicate that impacts will be unacceptable, special or additional management and contaminant control measures can be considered, including modification to the dredging operation or site; treatment of effluent, runoff, or leachate; treatment of dredged material solids; and site controls, such as surface covers or liners (USACE and EPA, 1992). Techniques for evaluating pathways have been developed (USACE and EPA, 1992; Myers et al., in press). Key considerations are summarized in Table 5-11.

The cost of using CDFs to contain contaminated sediments ranges from $15 to $50/yd^3, plus the operation and maintenance costs associated with closed CDFs (EPA, 1993a). Thus, storage in a CDF can be less expensive than landfill disposal, which can cost $20 to $120/yd^3 (EPA, 1994b). The design, construction, and operation of CDFs require conventional engineering approaches that have been used successfully for numerous other projects (USACE, 1987). A CDF can foster harbor development in urban areas; however, near-shore space may be difficult to find if wetlands must be consumed. In some cases it may be difficult to find an area and construct dikes in deep water to accommodate large volumes of material. If a freshwater CDF is located above an aquifer, controls may be required to prevent groundwater contamination or oxygenation of the sediment by rainfall, because the acids formed may cause the release of metals to groundwater.

Ex situ treatment usually requires a containment facility where the sediment is stored, dewatered, and pretreated (EPA, 1994a). Therefore, CDFs are often used in combination with pretreatment or more permanent treatment methods, a hybrid approach that offers the advantages of reducing, rather than simply transferring, contamination and fostering the reuse of storage space. Various processes are used to treat materials in CDFs. The pilot demonstration of bioremediation in a contained facility at Sheboygan River (cited earlier) is an example. A CDF can be similar to a bioreactor, which can be engineered to provide the conditions for stimulating microbial activity. CDFs can also be repositories for the natural degradation of contaminants. In studies of CDFs in Wisconsin and New York state, the USACE found that polyaromatic hydrocarbons appear to degrade in sunlight, suggesting that CDFs might be designed to advance natural processes by, for example, arranging for managed cycling of thin layers of sediments (T. Myers, USACE, personal communication to Marine Board staff, December 15, 1995).

TABLE 5-11 Confined Disposal Facility

State of Practice (system maturity, known pilot studies, etc.)	Applicability	Advantages/Effectiveness	Limitations	Research Needs
(a) The most commonly used placement alternative for contaminated sediments; (b) hundreds of sites nationwide for navigation dredging projects; (c) often used for pretreatment prior to final placement or as final sediment placement site for remediation projects.	Applicable to a wide variety of sediment types and project conditions.	(a) Low cost compared to ex situ treatment; (b) compatible with a variety of dredging techniques, especially direct placement by hydraulic pipeline; (c) proper design results in high retention of suspended sediments and associated contaminants; (d) engineering for basic containment normally involves conventional technology; (e) controls for contaminant pathways usually can be incorporated into site design and management; (f) conventional monitoring approaches can be used; (g) site can be used for beneficial purposes following closure, with proper safeguards.	(a) Does not destroy or detoxify contaminants unless combined with treatment; (b) control of some contaminant-loss pathways may be expensive.	(a) Design approaches, such as covers and liners, needed for low-cost contaminant controls; (b) design criteria for treatment of releases or control strategies for high-profile contaminants; (c) methods for site management to allow restoration of site capacity and potential use of treated materials.

133

The recovery of CDF space is practical for navigation dredging, and management guidelines are available (Montgomery et al., 1978). In some parts of the country, reusing CDF space may be more cost effective than constructing new facilities now that soil washing costs have dropped to approximately $20/yd^3 (J. Miller, USACE, personal communication to Marine Board staff, December 1, 1995). Reuse of the space may become increasingly common in both navigation and environmental dredging projects given rising construction costs and the difficulty of obtaining sites for new facilities. However, CDFs built in the 1970s were inexpensive (about $5/yd^3), and local officials must be convinced that reusing them can be more cost effective than expanding them.

Contained Aquatic Disposal

CAD involves the controlled placement of contaminated material at an open-water location, followed by covering with clean material. This method is similar in many respects to in situ capping except the CAD method involves relocating and containing the contaminated material laterally to minimize the spread of contamination across the bottom. With lateral containment, the volume of sediment needed for capping material is also minimized. Strategic placement can involve taking advantage of bottom depressions (either natural or excavated) or of target areas behind subaqueous dikes. Covered sediment can also form low-level mounds, with clean material spread above and beyond the edges of the contaminated pile. Figure 5-3 is an illustration of CAD. Key considerations are summarized in Table 5-12.

The CAD approach is particularly useful for disposing of contaminated dredged material. It is also applicable to contaminated sites in waters that are too shallow to permit in situ capping. The technique has been used in the Duwamish Waterway in Seattle (Sumeri, 1984; Truitt, 1986), in other countries (Averett and Francingues, 1994), and is planned for use in Boston Harbor (see Appendix C). To the committee's knowledge, CAD has not been used in any environmental cleanup projects.

The state of practice of CAD for restoring bottom sediment is not well advanced. Like in situ capping, a successful CAD operation requires only that the cap that isolates the contaminated material be accurately placed and well maintained. It is important that CAD be carried out in areas where erosion is minimal or controllable. The USACE has developed guidelines for planning CAD projects (Truitt, 1987a,b), determining the required capping thickness (Sturgis and Gunnison, 1988), determining design requirements (Palermo, 1991a), selecting sites (Palermo, 1991b), evaluating equipment and placement techniques (Palermo, 1991c), and evaluating monitoring systems (Palermo et al., 1992). In cooperation with the EPA, the USACE has also developed guidelines for in-place capping for restoration purposes (Palermo and Miller, 1995). A joint USACE and EPA technical

TABLE 5-12 Contained Aquatic Disposal

State of Practice (system maturity, known pilot studies, etc.)	Applicability	Advantages/Effectiveness	Limitations	Research Needs
Limited application. Reviews exist concerning (a) necessary data, equipment, and procedures; (b) engineering considerations; (c) guidelines for cap armoring design; (d) predicting chemical containment effectiveness.	(a) Costs and environmental effects of relocation are factors; (b) suitable types and quantities of cap material are available; (c) hydrologic conditions will not compromise the cap; (d) cap can be supported by original bed; (e) appropriate for sites where excavation is problematic or removal efficiency is low; (f) cap material is compatible with existing aquatic environment.	(a) Eliminates need to remove contaminated sediments; (b) cost effective for sites with large surface areas; (c) effective in containing contaminants by reducing bioaccessibility; (d) promotes in situ chemical or biological degradation; (e) maintains stable geochemical and geohydraulic conditions, minimizing contaminant release to surface water, groundwater, and air.	(a) Laboratory and field validation of capping procedures and tools; (b) analysis of data from existing and ongoing field demonstrations to support capping effectiveness; (c) test for chemical release during bed placement and consolidation; (d) tests to evaluate and simulate the effects of cap penetration by deep burrowing organisms; (e) simulate and evaluate consequences of mixing; (f) potential loss of contaminants to the water column may require controls during placement.	(a) Design criteria for treatment of releases or control strategies for high-profile contaminants; (b) improved methods for evaluation of potential contaminant release pathways; (c) develop reliable cost estimates.

document for the subaqueous capping of dredged material is also in preparation (Palermo et al., in press).

A major advantage of CAD is that it can be performed with conventional dredging equipment, although the equipment may have to be operated in special ways. Also, unlike the CDF option, the chemical environment surrounding the contaminated material remains virtually unchanged because the sediment remains in the waters of its origin. A major consideration is the potential loss of contaminated sediments during placement operations. Controls comparable to the ones used with CDF technology must be applied to minimize such losses. Research is needed to improve control capabilities and to determine the effects of losses on the ecosystem and to assess the associated risks. Research on the long-term effectiveness of various types of capping, including CAD, is also needed. Resolution of these issues would probably enhance the acceptability of this technology for restoring contaminated sediment sites. The committee could not locate any useful data on the actual costs of CAD.

Another possible approach to subaqueous offshore containment, at least for small volumes of material, might be to encase contaminated sediments in woven or nonwoven permeable synthetic fabrics. The casings could be expected to eliminate losses during placement and to contain the contaminated sediment on the seafloor. Fabric has been used for some 30 years to make various types of receptacles, such as sandbags, geotextile tubing, and geotextile containers (see Fowler et al., 1994; Pilarczyk, 1994, and references therein). This approach was demonstrated with contaminated materials dredged from Marina del Ray in California, where the use of geotextile containers added more than $50/yd^3 to the cost of the project (Clausner, 1996). Because most contaminants are sorbed to sediments and would not seep through the fabric, placement of filled geotextile bags in the water might be environmentally safe and would eliminate the need for land-based disposal sites. However, no data are available about the environmental effects of this approach (Clausner, 1996). A collection of bags could be capped, if necessary.

In addition to their utility in civil engineering projects and in the dewatering of dredged sediments, geotextile containers could provide a unique system for demonstrating emerging ex situ bioremediation technologies for certain contaminants. As disposal sites become increasingly difficult to find, the treatment of contaminated sediments in constructed cells, CDFs, or geotextile containers could be ways of reusing scarce sites.

Another idea that has received some attention is the placement of contained wastes on the abyssal plain (roughly 4,500 m deep) in the ocean. This idea was recently examined in a U.S. Department of Defense-sponsored study of ways to place and monitor clean dredged material, sewage sludge, and combustion fly ash (Valent and Young, 1995). The most attractive technique involves the use of fabric-like containers to isolate wastes from the water column during deployment from the transport ship or barge. Although this proposal has technical merit, legal barriers (dumping of contaminated sediments in the open ocean is not allowed) and environmental uncertainties must still be investigated. And because of the expense of the long-distance ocean transport of contaminated sediment, the cost-effectiveness of this idea needs to be closely examined.

Landfill Disposal

Contaminated dredged material sometimes is placed in licensed solid-waste landfills. Dredged material has also been used on a limited basis as solid-waste landfill cover. Placement in landfills may be an affordable and timely disposal option, especially for small volumes of contaminated material. Treated dredged material from remediation projects has been placed in landfills for nonhazardous solid waste, sometimes at great distances from the remediation site. For example, treated sediment from the Marathon Battery Superfund site in New York was transported to a landfill in Michigan (see Appendix C), and the Record of Decision (ROD) for the United Heckathorn Superfund site in California calls for placing sediment in a landfill in Utah (Palermo, 1995). Key considerations are summarized in Table 5-13.

In some ways, landfill disposal and the containment of contaminated sediments are similar to the methods used for handling municipal and conventional hazardous waste. However, handling sediment differs dramatically from conventional landfill operations because the contaminated sediments usually have a high water content or are in slurry form. Solid-waste landfills cannot accept free liquids so the sediments must be dewatered. Use of a CDF as a pretreatment facility for dewatering sediments, or mechanical dewatering and possibly stabilization, are steps that can be taken prior to transporting sediment to a landfill. Another factor limiting landfill placement is that licensed landfills in most regions of the country do not have the capacity to accommodate large volumes of additional material.

EVALUATING THE PERFORMANCE OF TECHNOLOGIES AND CONTROLS

The performance of sediment management technologies and controls must be evaluated for every project, not only to determine if specific objectives have been met but also to gather data for improving the state of the art. Monitoring is the principal method of evaluating performance. The subject of monitoring sites targeted for remediation was addressed previously by the NRC (1990). However, the committee wishes to emphasize the importance of performance evaluation and to point out several ways current approaches might be improved. Three topics are discussed: interim controls, long-term monitoring, and cost-benefit analysis.

TABLE 5-13 Landfills

State of Practice (system maturity, known pilot studies, etc.)	Applicability	Advantages/Effectiveness	Limitations	Research Needs
Used for several dredged material and Superfund projects involving contaminated sediments.	(a) Small volumes; (b) where no other alternatives or sites are available.	(a) Does not require acquisition of permanent placement site; (b) may be most cost effective for small volumes; (c) effectiveness is inherent in the site license.	(a) Lack of landfill capacity in most regions of the country; (b) requires handling and transport to the landfill; (c) restriction on free liquids requires dewatering as a pretreatment step.	Improved methods for rehandling, dewatering, and transporting dredged sediments.

Evaluation of Interim Controls

Little is known about the effectiveness of interim controls. Administrative measures, such as the surveys cited earlier that focused on fishermen's attitudes (Belton et al., 1985), would be helpful. The direct observation of structural controls, such as the approach used at Manistique Harbor in Michigan (a project discussed earlier in this chapter), would not only provide physical evidence of performance but could also be designed to evaluate risk reduction. To be most useful, monitoring should be done with an eye toward improving the future application of interim controls.

Long-Term Monitoring

Monitoring can involve physical, chemical, biological, or toxicological processes, or combinations thereof. Monitoring needs to have a specific purpose and must be tailored to the specific remediation process or technology.

For example, monitoring in situ and ex situ containment systems needs to include physical assessment of the barrier, chemical analysis of contaminant mobility, and, perhaps, measurement of biological characteristics. Monitoring during treatment must take place at appropriate intervals. Systems for monitoring incinerator performance, for example, respond to upsets in minutes. Given the expense and time required to measure metal, PCB, and polyaromatic hydrocarbon concentrations, monitoring systems usually measure other aspects of process performance, such as measuring for the presence of carbon monoxide in effluent gases as an indicator of incomplete combustion. The monitoring of treatments that involve the repeated washing of sediments in batches may require chemical analysis of each batch of sediment.

Because the utility of predictions provided by numerical models is limited by uncertainties and gaps in the scientific and technical knowledge, field monitoring of contaminated sites is needed before, during, and after remediation. Monitoring may provide surprising results, so the site management structure must be sufficiently flexible to respond to new information or unexpected events.

The committee's major concern about monitoring is the apparent asymmetry in the current state of practice. Initial site assessments to define contamination levels and distribution are carried out with great precision. But post-project monitoring tends to be more qualitative than quantitative. In most cases, no effort is made to examine directly whether specific, risk-based objectives were actually met. Risk-based monitoring could not only improve the rigor of project evaluation but could also provide data for the calibration of methods for predicting success.

Cost-Effectiveness Analysis

It is extremely difficult to evaluate the costs associated with remediation technologies because the data are not collected in a uniform manner. Available data are inconsistent with respect to both the types of costs included and the units of measure (e.g., cubic yards, tons, hectares). Geographical variations are not usually considered. The problem stems partly from the lack of a formal structure for reporting cost data. Even if good cost data were available, improved methods of measuring effectiveness would be needed for reliable comparative analyses of technologies on the basis of cost effectiveness. But post-project monitoring tends to be qualitative rather than quantitative.

Although the available cost data are limited, they are sufficient for estimating cost ranges for various remediation technologies. The costs of removing and transporting contaminated sediments (generally less than \$15 to \$20/yd^3) tend to be higher than the costs of conventional dredging (seldom more than \$5/yd^3) but much lower than the costs of ex situ treatment (which can cost well over \$100/yd^3 and sometimes more than \$1,000/yd^3). For systems involving precision dredging technology, there is a potential for reducing costs still further. Volume reduction (i.e., removing only those sediments that require treatment and entraining as little water as possible) can mean greater cost savings than increased dredging rates. When the volume of contaminated sediment exceeds 10,000 yd^3, total treatment costs can be appreciable, but economy of scale reduces the unit cost.

Treatment costs can also be reduced through pretreatment to separate contaminated silt- and clay-sized particles from generally cleaner sand; however, the cost of this process (\$20 to \$50/yd^3) is generally justified only if there is a large proportion of sand. The costs of in situ treatments could be less than \$100/yd^3, but in situ approaches have not been demonstrated. Given the chemical complexity of the waste mixtures, it is likely that a sequence of treatment processes will be required.

It is important to emphasize that the absence of detailed, reliable cost data for many remediation technologies does not pose a major barrier to project planning because the unique conditions (geographical and otherwise) of each situation demand that costs always be estimated individually for each case. However, improved reporting of cost information for full-scale remediation systems would permit fair, overall comparisons and would provide benchmarks for future R&D and systems design. The collection of reliable, standardized cost data would help decision makers quickly choose technologies that could be effective at a particular site within a given budget. The need for standardized cost data for environmental cleanup projects in general has been recognized by some government[9]

---

[9] A federal interagency cost estimating group, which includes representatives from the EPA and USACE, has been formed (Rubin, 1995).

and industrial leaders, who are collaborating to develop a uniform approach (Rubin, 1995).

RESEARCH, DEVELOPMENT, TESTING, AND DEMONSTRATION

Needs for R&D, testing, and demonstration programs have been identified throughout this chapter. Specifically:

- Few data are available on the use or effectiveness of interim control technologies, and some promising approaches, such as using CDFs for the temporary storage and treatment of contaminated sediments, have yet to be developed fully.
- The use of in situ technologies is limited by a lack of understanding of the fundamental processes of the transport, degradation, and biological accumulation of contaminants under both natural and engineered conditions, coupled with the difficulty of implementation and process control in extremely variable and complex natural environments.
- The United States has little experience with environmental dredging because the approach is fairly new. Some specialized sediment removal systems are available for unusual site conditions and, through demonstration programs, they could be applied in a wider range of circumstances. But the advantages of specialized equipment as compared with conventional dredges must be documented through direct field comparisons. Advances in the precision and accuracy of dredging can be applied widely and make sense as long as they are consistent with the level of definition of the vertical and horizontal extent of contamination.
- The implementation of CDFs and CAD could be improved. Design criteria for the control of contaminant release pathways, low-cost treatment options, and management approaches to permit the reuse of storage capacity are needed, as is the development of potential beneficial uses for treated material. The use of CAD requires improved tools for the designing and monitoring of sediment caps and armor layers and for evaluating cap placement and the long-term stability of caps and their effectiveness in isolating contaminants.
- Ex situ treatment technologies are still at an early stage of development. The costs of these processes need to be reduced. In addition, these technologies need to be evaluated with respect to their effectiveness in reducing environmental exposures from contaminants released to air and water, as well as from contaminants that remain in the sediments.

The importance of attending to each stage of the technology development process cannot be overemphasized. The process spans five phases: concept, bench scale, pilot scale, demonstration (field scale), and commercialization. Very little work with marine sediments has gone beyond the bench-scale stage, where theories and empirical experience are tested in the laboratory. Because of the unique characteristics of any site and the lack of experience with many sediment handling and remediation technologies, bench-scale and pilot-scale tests, as well as demonstration projects, are needed prior to the full-scale implementation of innovative approaches. The success of the ARCS research and planning program in the Great Lakes can be attributed, in part, to the emphasis on technology demonstration, as well as to the scientific rigor imposed by peer review of proposed methods. Another program is under way for remediating New York Harbor sediments that will bench test, pilot test, and demonstrate treatment technologies (Stern et al., 1994), but more work is needed.

There is a particular need for side-by-side comparisons of innovative and conventional dredging and remediation technologies so that developers' claims can be evaluated and verified. At present, there is no formal, unbiased mechanism for identifying and evaluating emerging technologies, and new ideas are transferred to the field very slowly. In the United States, detailed demonstrations and comparisons of sediment-handling and remediation technologies have been limited to the ARCS program and the EPA's Superfund Innovative Technology Evaluations (SITE) program, in which manufacturers pay for the demonstration of new technologies for the cleanup of toxic and hazardous waste sites. Following the SITE program model, a mechanism could be established for making unbiased technical evaluations of innovative sediment-handling and remediation technologies based on real-time, realistic project conditions. The program could arrange for side-by-side demonstrations of innovative and conventional technologies at suitable sites under strict protocols for technical and economic evaluations.

COMPARATIVE ANALYSIS OF TECHNOLOGY CATEGORIES

The committee considered various ways of summarizing its evaluation of remediation technologies and ultimately settled on a qualitative comparison based on key attributes. Table 5-14 provides the foundation for the comparison by summarizing the state of practice for the general technology categories, using information provided in this chapter. Building on this information, Table 5-15 displays the committee's overall assessment using the criteria identified in the statement of task. This section discusses Table 5-15, which was developed by the committee based on the analysis in this report and the experience and expertise of individual committee members.

The column on effectiveness is an order-of-magnitude estimate of contaminant reduction or isolation and removal efficiency; the score is roughly equivalent to the total number of 9s in the removal efficiency (e.g., a score of 3 is three 9s or 99.9 percent removal efficiency). The feasibility column represents the extent of technology development. The lowest score means a concept has not been verified experimentally; the next-lowest score means a technology has been

TABLE 5-14 Qualitative Comparison of the State of the Art in Remediation Technologies[a]

Feature technology	State-of-design Guidance	Number of Times Used	Scale of Application	Cost (per cubic yard)	Limitations[b]
Natural recovery	Nonexistent	2	Full scale	Low	Source control Sedimentation Storms
In-place containment	Developing rapidly	< 10	Full scale	< $20	Limited technical guidance Legal/regulation uncertainty
In-place treatment	Nonexistent	≈2	Pilot scale	Unknown	Technical problems Few proponents Need to treat entire volume
Excavation and containment	Substantial and well developed	Several hundred	Full scale	$20 to $100	Site availability Public assistance
Excavation and treatment	Limited and extrapolated from soil	< 10	Full scale	$50 to $1,000	High cost Inefficient for low concentration Residue toxic Need for treatment train

[a]Estimates for North America.
[b]See Table 5-15 for further details.

TABLE 5-15 Comparative Analysis of Technology Categories

Approach	Feasibility	Effective	Practicality	Cost
INTERIM CONTROL				
Administrative	0	4	2	4
Technological	1	3	1	3
LONG-TERM CONTROL				
In Situ				
Natural recovery	0	4	1	4
Capping	2	3	3	3
Treatment	1	1	2	2
Sediment Removal and Transport	2	4	3	2
Ex Situ Treatment				
Physical	1	4	4	1
Chemical	1	2	4	1
Thermal	4	4	3	0
Biological	0	1	4	1
Ex Situ Containment	2	4	2	2
SCORING				
0	< 90%	Concept	Not acceptable, very uncertain	$1,000/yd^3
1	90%	Bench		$100/yd^3
2	99%	Pilot		$10/yd^3
3	99.9%	Field		$1/yd^3
4	99.99%	Commercial	Acceptable, certain	< $1/yd^3

demonstrated at the bench level in a small (typically a batch) reactor. Higher scores represent, in ascending order, a pilot-scale demonstration using contaminated sediments in a volume on the order of a few cubic yards, a field-scale demonstration using tens of cubic yards, and finally, a commercial operation. The practicality ranking reflects public acceptance; a score of 0 means the public would not tolerate such an activity, and a score of 4 means a technology would be viewed favorably. The practicality ranking also includes some qualitative measure of uncertainty, which can be a deciding factor to a risk-averse regulatory community and public. Finally, the cost score is inversely related to the treatment cost, with incineration being the most expensive and thus assigned the lowest score. Costs do not include expenses associated with monitoring, environmental resource damages, or the costs imposed on the public by closure of a commercial fishery or loss of subsistence fishing.

In the category of interim controls, two approaches were considered: administrative controls that provide warnings and structural controls that isolate contaminated sediments from humans and ecosystems. Administrative controls, such as the controls used during the natural restoration of the James River estuary, are probably less than 90 percent effective in limiting human consumption of finfish and shellfish contaminated by sediments. Administrative controls would be most effective in restricting commercial operations and least effective in limiting subsistence fishing, particularly fishing by individuals unable to read posted signs. Administrative controls do not limit ecosystem exposures unless measures are taken to exclude wildlife from contaminated areas. Administrative controls appear to be practical although the public perceives that the responsible parties are doing nothing besides posting signs. The costs of administrative interim controls are very low, but there is some uncertainty as to the type and level of monitoring program that would be required.

Technology-based interim controls have the potential to effectively limit contaminant releases to the ecosystem, although there has been little experience with this approach. The practicality score is low because of concerns that the contamination will not be remediated completely. The cost is relatively low, but it can rise if extensive monitoring, which may last indefinitely, is required. The potential exists for cost savings if the interim control becomes the long-term control, but there is an alternative risk of increasing costs in the future if the interim control has to be removed. In the latter case (e.g., if removal of a cap resulted in the mixing of clean and contaminated sediment), the project might entail the removal and treatment of larger volumes of diluted, contaminated sediments than were present originally.

Although in situ controls are attractive in some ways, there is considerable doubt about their effectiveness and practicality. Natural recovery is of limited effectiveness in preventing contaminant release into the ecosystem, because this approach depends on natural processes of burial by sedimentation and contaminant destruction or sequestration by physical, chemical, or microbial processes. Natural recovery was demonstrated at the James River. The cost borne by the responsible party and the regulatory community is low.

In situ control by in-place capping involves a number of trade-offs compared with natural recovery. Laboratory experiments and calculations based on chemical and physical principles indicate that capping should be at least 99 percent effective in reducing contaminant release over the long term. The technology has been demonstrated at the field scale, although long-term performance has not been verified. Some stakeholders view capping as a temporary solution and thus of less-than-optimum practicality. Costs, including monitoring, are moderate.

In situ treatment using physical, chemical, and biological approaches is at an early stage of development and testing. Limited information is available on the effectiveness of these processes because most studies have not gone beyond the bench scale. Given the limited experience and the uncertainties about effectiveness and cost, in situ treatment may seldom be acceptable to risk-averse decision makers and stakeholders.

The next category, sediment removal and transport, is the first step in ex situ remediation. There is an extensive U.S. commercial experience base for this technology with navigation dredging and the placement of dredged material. Sediments can be recovered and isolated with contaminant losses of approximately 2 to 5 percent. Experience with clean sediments provides reasonable certainty regarding the feasibility and cost, although the practicality of dredging is often not completely accepted by the public, particularly when contaminated sediments are involved. Costs are moderate for environmental dredging and for transport.

A wide array of ex situ technologies has been considered. Four general treatment categories and one containment technology are listed in Table 5-15. These approaches are feasible and practical although they are costly, and few have been demonstrated at pilot or full scale. Physical treatment methods separate sediments based on size and density. The approach is commercially feasible in large-scale mining operations and has been used in the management of contaminated sediments. The effectiveness of physical separation can be on the order of 90 percent if the contaminants selectively associate with a small mass fraction of the sediments that can be isolated; further treatment of the concentrated contaminants is then required. Costs are moderate.

Ex situ chemical treatments are less well developed than physical separation technologies. The effectiveness rating is low because results to date at the bench and pilot scales show only 90 percent recovery of contaminants. For sediments contaminated by both organics and metals, even lower recoveries can be expected, and multiple treatment processes need to be sequenced. Because full-scale experience with contaminated sediments is limited, the feasibility score of chemical treatments is also low.

Thermal technologies have the highest effectiveness of any remediation technology, with the capability of destroying more than 99.99 percent of organic contaminants, including PCBs. There has been considerable commercial experience in destroying hazardous waste by incineration, and the regulatory community and most stakeholders understand the principles of this approach. But there is still some skepticism about the technology. The major drawback to thermal destruction is high cost, which can reach $1,000/yd^3 at low processing rates.

Ex situ biological treatment approaches have some potential, and the concept is supported by most stakeholders. However, few data are available on effectiveness, and studies have been limited to the bench scale. Much of the expertise evolving with the biological remediation of soils and groundwater can be applied to sediments, but additional research is needed to adapt to the unique contaminant mixtures, the saltwater content, and the fined-grained nature of marine sediments. In addition, knowledge is limited concerning the effects of contaminant mixtures, particularly mixtures of organics and metals, on biological processes.

The containment of residues in a facility above or under the water is a common sediment management technique, so there is a record of performance. Containment systems are effective in containing at least 99 percent of the contaminants initially and can provide long-term isolation if the physical integrity of the container is maintained. The major downside to this approach is the difficulty of finding sites for the facilities and gaining public acceptance of a landfill for sediments. The costs are low to moderate.

Of most interest to the committee is the obvious need to make trade-offs in the selection of technologies. Interim controls and in situ approaches are both feasible and relatively inexpensive but limited in terms of effectiveness, practicality, and uncertainty. Ex situ approaches require sediment removal and transport, which receive high scores, combined with treatment and containment approaches, which receive good scores for feasibility and practicality but low scores for effectiveness and cost. Thus, the decision maker is left in the uncomfortable position of trading off low-cost, less-effective, less-practical, yet feasible interim controls and in situ approaches, as compared with the most practical ex situ approaches, which can be effective but tend to be expensive and complex. The magnitude of the contamination problem and site-specific considerations can guide the decision maker in analyzing these alternatives. One solution to this dilemma can be found through cost-benefit analysis (see Chapter 2), a decision tool that uses remediation technologies as one of several inputs.

In comparing the results of the qualitative assessment with the history of use (Table 5-14), it appears that feasibility and practicality are the most important considerations in the implementation of technologies or controls and that high cost is a serious disincentive.

REFERENCES

Abramowicz, D.A., M.R. Harkness, J.B. McDermott, and J.J. Salvo. 1992. 1991 In situ Hudson River Research Study: A Field Study on Biodegradation of PCBs in Hudson River sediments. Schenectady, New York: General Electric Corporation Research and Development.

Allen, J.P. In press. Mineral Processing Pre-treatment of Contaminated Sediment. Great Lakes National Program Office. Chicago: EPA.

Alluvial Mining Group, Ltd. 1993. Tramrod with Environmental Dredging Tools. Operations Manual. Document No. RBW-AM-931. Sudbury, Suffolk, United Kingdom: Alluvial Mining Group, Ltd.

Averett, D.E., and N.R. Francingues. 1994. Sediment remediation: An international review. Pp. 596–605 in Dredging '94: Proceedings of the 2nd International Conference on Dredging and Dredged Material Placement. E.C. McNair, Jr., ed. New York: American Society of Civil Engineers.

Averett, D.E., B.D. Perry, E.J. Torrey, and J.A. Miller. 1990. Review of removal, containment and treatment technologies for remediation of contaminated sediment in the Great Lakes. Miscellaneous Paper EL-90-25. Vicksburg, Mississippi: U.S. Army Engineer Waterways Experiment Station.

Belton, T., B. Ruppel, K. Lockwood, S. Shiboski, G. Bukowski, R. Roundy, N. Weinstein, D. Wilson, and H. Whelan. 1985. A Study of Toxic Hazards to Urban Recreational Fishermen and Crabbers. New Jersey Department of Environmental Protection, Office of Science and Research, and Cook College–Rutgers University, Department of Human Ecology. September 15.

Bohlen, W.F. 1978. Factors governing the distribution of dredge resuspended sediments. Pp. 2001–2019 in Proceedings of the 16th Coastal Engineering Research Conference held August 17–September 3, 1978, in Hamburg, Germany. New York: American Society of Civil Engineers.

Bragg J.R., R.C. Prince, E.J. Harner, and R.M. Atlas. 1994. Effectiveness of bioremediation for the Exxon Valdez oil spill. Nature 368:413–418.

Buchberger, C. 1992. Environment Canada tests cable arm bucket on contaminated sediment in Toronto. International Dredging Review 11(5):6–7.

Buchberger, C. 1993. Environment Canada demonstrations: remediation technologies for the removal of contaminated sediments in the Great Lakes. Terra et Aqua, International Journal on Public Works, Ports, and Waterways Developments (50):3–13.

Caulfield, D.D., A. Ostaszewski, and J. Filkens. 1995. Precision Digital Hydroacoustic Sediment Characterization: Analysis in the Trenton Channel of the Detroit River. Paper presented at the 1995 Conference on Great Lakes Research. East Lansing: Michigan State University.

Clausner, J.E. 1996. Potential Application of Geosynthetic Fabric Containers for Open Water Placement of Contaminated Dredged Material. Technical Note EEDP-01-39. Vicksburg, Mississippi: U.S. Army Engineer Waterways Experiment Station.

Clausner, J.E., W.A. Birkemeier, and G.R. Clark. 1986. Field Comparison for Four Nearshore Survey Systems. Miscellaneous Paper CERC-86-6. Vicksburg, Mississippi: U.S. Army Engineer Waterways Experiment Station.

Collins, M.A. 1995. Dredging-Induced Near-Field Resuspended Sediment Concentrations and Source Strengths. Miscellaneous Paper D-95-2. Vicksburg, Mississippi: U.S. Army Engineer Waterways Experiment Station.

Crockett, T.R. 1993. Modeling Near Field Sediment Resuspension in Cutterhead Suction Dredging Operations. Master's thesis. Lincoln: University of Nebraska.

Cundy, D.F., and W.F. Bohlen. 1982. A numerical simulation of the dispersion of sediments suspended by estuarine dredging operations. Pp. 339–352 in Estuarine and Wetlands Processes. P. Hamilton and K.B. MacDonald, eds. New York: Plenum.

Detzner, H.D. 1993. Mechanical treatment of the dredged material from Hamburg Harbor. Pp. 3.25–3.28 in Proceedings of the CATS II Congress 1993. Antwerp, Belgium: Technological Institute of the Royal Flemish Society of Engineers.

Digiano, F.A., C.T. Miller, and J. Yoon. 1993. Predicting release of PCBs at point of dredging. Journal of Environmental Engineering 119(1):72–89.

Digiano, F.A., C.T. Miller, and J. Yoon. 1995. Dredging Elutriate Test (DRET) Development. Contract Report D-95-1. Vicksburg, Mississippi: U.S. Army Engineer Waterways Experiment Station.

Edgar, C.E., and R.M. Engler. 1984. The London Dumping Convention and its role in regulating dredged material: An update. Pp. 240–249 in Dredging and Dredged Material Disposal, ASCE Specialty Conference Dredging 1984, vol. 1. New York: American Society of Civil Engineers.

Environmental Protection Agency (EPA). 1989. Solidification and stabilization of CERCLA and RCRA wastes. Washington, D.C.: EPA.

EPA. 1990. Record of Decision, New Bedford Harbor Pilot Dredging Project. Boston: EPA, Region 1.

EPA. 1991. Handbook—Remediation of Contaminated Sediments. EPA-625/6-91-028. Center for Environmental Research Information, Office of Research and Development. Cincinatti: EPA.

EPA. 1993a. Selecting Remediation Techniques for Contaminated Sediment. Office of Water. EPA-823-B93-001. Washington, D.C.: EPA.

EPA. 1993b. Remediation Technologies Screening Matrix and Reference Guide. Office of Solid Waste and Emergency Response. EPA-542-B-93-005. Washington, D.C.: EPA.

EPA. 1994a. Assessment and Remediation of Contaminated Sediments (ARCS) Program, Final Summary Report. Great Lakes National Program Office. EPA-905-S-94-001. Chicago: EPA.

EPA. 1994b. Assessment and Remediation of Contaminated Sediments (ARCS) Program. Remediation Guidance Document. Great Lakes National Program Office. EPA 905-R94-003. Chicago: EPA.

EPA. 1994c. Superfund Record of Decision: Sangamo Weston/Twelve-Mile Creek/Lake Hartwell Site, Pickens, Georgia. Office of Emergency and Remedial Response. EPA/ROD/R04-94/178. Washington D.C.: EPA.

Fowler, J., D.J. Sprague, and D. Toups. 1994. Dredged Material-Filled Geotextile Containers, Environmental Effects of Dredging. Technical Notes. Vicksburg, Mississippi: U.S. Army Engineer Waterways Experiment Station.

Frodge, S.L., B.W. Remondi, and D. Lapucha. 1994. Dredging Research Technical Notes, Real-Time Testing and Demonstration of the U.S. Army Corps of Engineers' Real-Time On-The-Fly Positioning System. DRP-4-10. Vicksburg, Mississippi: U.S. Army Engineer Waterways Experiment Station.

Galloway, J.E., and F.L. Snitz. 1994. Pilot-scale demonstration of sediment washings. Pp. 981–990 in Dredging '94: Proceedings of the 2nd International Conference on Dredging and Dredged Material Placement. E.C. McNair, Jr., ed. New York: American Society of Civil Engineers.

Garbaciak, S. 1994. Laboratory and field demonstrations of sediment technologies by the U.S. EPA's Assessment and Remediation of Contaminated Sediments (ARCS) Program. Pp. 567–578 in Dredging '94: Proceedings of the 2nd International Conference on Dredging and Dredged Material Placement. E.C. McNair, Jr., ed. New York: American Society of Civil Engineers.

Gibbs, R.J. 1973. Mechanisms of trace metal transport in rivers. Science 180:71–73.

Hahnenberg, J. 1995. Presentation at the Workshop on Interim Controls held July 31, 1995. Committee on Contaminated Sediments, National Research Council. Chicago: EPA Headquarters.

Harkness, M.R., J.B. McDermott, D.A. Abramowicz, J.J. Salvo, W.P. Flanagan, M.L. Stephens, F.J. Mondello, R.J. May, and J.H. Lobos. 1993. In situ stimulation of aerobic PCB biodegradation in Hudson River sediments. Science 259:503–507.

Hayes, D.F. 1993. Assessing impacts of environmental dredging operations. Pp. 161–172 in Proceedings of the 16th U.S./Japan Experts Meeting on Management of Bottom Sediments Containing Toxic Substances held October 12–14, 1993, in Kitakyushu, Japan. Unpublished.

Hayes, D.F., N. McLellan, and C.L. Truitt. 1988. Demonstrations of Innovative and Conventional Dredging Equipment at Calumet Harbor, Illinois. Miscellaneous Paper EL-88-1. Vicksburg, Mississippi: U.S. Army Engineer Waterways Experiment Station.

Herbich, J.B. 1995. Removal of contaminated sediments: Equipment and recent field studies. Pp. 77–111 in Dredging, Remediation, and Containment of Contaminated Sediments. K.R. Demars, G.N. Richardson, R.N. Yong, and R.C. Chaney, eds. ASTM STP 1293. Philadelphia: American Society of Testing and Materials.

Herbich, J.B., and J. DeVries. 1986. An Evaluation of the Effects of Operational Parameters on Sediment Resuspension During Cutterhead Dredging Using a Laboratory Model Dredge System. Report No. CDS 286. College Station: Texas A&M University.

Herbich, J.B., and S.B. Brahme. 1991. Literature Review and Technical Evaluation of Sediment Resuspension During Dredging. Contract Report HL-91-1. Prepared for the U.S. Army Engineer Waterways Experiment Station, Vicksburg, Mississippi.

Hoigné, J. 1988. The chemistry of ozone in water. Pp. 121–143 in Process Technologies for Water Treatment. S. Stucki, ed. New York: Plenum.

Hua, I., R.H. Hochemer, and M.R. Hoffmann. 1995. Sonochemical degradation of p-nitrophenol in a parallel-plate near-field acoustical processor. Environmental Science and Technology 29(11):2790–2796.

Huggett, R.J., and M.E. Bender. 1980. Kepone in the James River. Environmental Science and Technology 14(8):918–923.

Kato, H. 1993. Development of thin-layer dredging equipment with belt conveyor. Pp. 173–190 in Proceedings of the 16th U.S./Japan Experts Meeting on Management of Bottom Sediments Containing Toxic Substances held October 12–14, 1993, in Kitakyushu, Japan. Unpublished.

Keillor, J.P. 1993. Obstacles to the remediation of contaminated soils and sediments in North America at reasonable cost. In Proceedings of the CATS II Congress: Characterization and Treatment of

Contaminated Dredged Material. Antwerp, Belgium: Technological Institute of the Royal Flemish Society of Engineers.

Kenna, B.T., S.M. Yaksich, D.E. Averett, and M.A. Zappi. 1994. Demonstration of equipment for dredging contaminated sediments at Buffalo River, Buffalo, New York. Pp. 885–895 in Dredging '94: Proceedings of the 2nd International Conference on Dredging and Material Placement. E.C. McNair, Jr., ed. New York: American Society of Civil Engineers.

Krahn, H.P. 1990. Feasibility Study of Estuary and Lower Harbor Bay, New Bedford, Massachusetts, vol. 2. EBASCO Services, Inc.

McGee, R.G., R.F. Ballard, Jr., and D.D. Caulfield. 1995. A Technique to Assess the Characteristics of Bottom and Subbottom Marine Sediments. Technical Report DRP-95-3. Vicksburg, Mississippi: U.S. Army Engineer Waterways Experiment Station.

McLellan, T.N., R.N. Havis, D.F. Hayes, and G.L. Raymond. 1989. Field Studies of Sediment Resuspension Characteristics of Selected Dredges. Technical Report HL-89-9. Vicksburg, Mississippi: U.S. Army Engineer Waterways Experiment Station.

Miller, J. 1995. Presentation at the Workshop on Interim Controls held July 31, 1995. Committee on Contaminated Sediments, National Research Council. Chicago: EPA Headquarters.

Montgomery, R.L., A.W. Ford, M.E. Poindexter, and M.J. Bartos. 1978. Guidelines for Dredged Material Disposal Area Reuse Management. Technical Report DS-78-12. Vicksburg, Mississippi: U.S. Army Engineer Waterways Experiment Station.

Moore, J.N., E.J. Brook, and C. Johns. 1989. Grain size partitioning of metals in contaminated coarse-grained river flood plain sediment, Clark Fork River, Montana, USA. Environmental Geology and Water Science 14(2):107–115.

Myers, T.E., and M.E. Zappi. 1992. Laboratory evaluation of stabilization/solidification technology for reducing the mobility of heavy metals in New Bedford Harbor Superfund site sediment. Pp. 304–319 in Stabilization and Solidification of Hazardous, Radioactive, and Mixed Wastes, vol. 2. T.M. Gilliam and C.C. Wiles, eds. ASTM STP 1123. Philadelphia: American Society of Testing and Materials.

Myers, T.E., M.R. Palermo, T.J. Olin, D.E. Averett, D.D. Reible, J.L. Martin, and S.C. McCutcheon. In press. Estimating Contaminant Losses from Components of Remediation Alternatives for Contaminated Sediments. Great Lakes National Program Office. Report prepared for U.S. Environmental Protection Agency, Chicago, Illinois.

Nakles, D.V., and D.G. Linz, eds. In press. Environmentally Acceptable Endpoints in Soil. Annapolis, Maryland: American Academy of Environmental Engineers.

National Research Council (NRC). 1989. Contaminated Marine Sediments: Assessment and Remediation. Washington, D.C.: National Academy Press.

NRC. 1990. Managing Troubled Waters: The Role of Marine Environmental Monitoring. Washington, D.C.: National Academy Press.

Otis, M.J. 1992. A pilot study of dredging and disposal alternatives for the New Bedford Harbor, Massachusetts, Superfund site. In Proceedings of the 14th U.S./Japan Experts Meeting on Management of Bottom Sediments Containing Toxic Substances in Yokohama, Japan. T.R. Patin, ed. Vicksburg, Mississippi: U.S. Army Engineer Waterways Experiment Station.

Otis, M.J. 1994. New Bedford Harbor, Massachusetts, dredging/disposal of PCB-contaminated sediments. Pp. 579–595 in Dredging '94: Proceedings of the 2nd International Conference on Dredging and Material Placement. E.C. McNair, Jr., ed. New York: American Society of Civil Engineers.

Palermo, M.R. 1991a. Design Requirements for Capping. Dredging Research Technical Notes, DRP-05-03. Vicksburg, Mississippi: U.S. Army Engineer Waterways Experiment Station.

Palermo, M.R. 1991b. Site Selection Considerations for Capping. Dredging Research Technical Notes, DRP-5-04. Vicksburg, Mississippi: U.S. Army Engineer Waterways Experiment Station.

Palermo, M.R. 1991c. Equipment and Placement Techniques for Capping. Dredging Research Technical Notes, DRP-5-05. Vicksburg, Mississippi: U.S. Army Engineer Waterways Experiment Station.

Palermo, M.R. 1995. Considerations for disposal of dredged sediments in solid waste landfills. Paper prepared for the 16th Technical Conference of the Western Dredging Association and the 28th Annual Texas A&M Dredging Seminar and University of Wisconsin Sea Grant Dredging Workshop held May 23–26, 1995, in Minneapolis, Minnesota. Available from M.R. Palermo, U.S. Army Engineer Waterways Experiment Station, Vicksburg, Mississippi.

Palermo, M.R., and J. Miller. 1995. Strategies for management of contaminated sediments. Pp. 289–296 in Dredging, Remediation, and Containment of Contaminated Sediments. K.R. Demars, G.N. Richardson, R.N. Yong, and R.C. Chaney, eds. ASTM STP 1293. Philadelphia: American Society of Testing and Materials.

Palermo, M.R., T.J. Fredette, and R.E. Randall. 1992. Monitoring Considerations for Capping. Dredging Research Technical Notes, DRP-05-07. Vicksburg, Mississippi: U.S. Army Engineer Waterways Experiment Station.

Palermo, M.R., R.M. Engler, and N.R. Francingues. 1993. The U.S. Army Corps of Engineers perspective on environmental dredging. Buffalo Environmental Law Journal 1(2):243–253.

Palermo, M.R., R.E. Randall, T. Fredfette, and J. Clausner. In press. Technical Guidance for Subaqueous Dredged Material Capping. Vicksburg, Mississippi: U.S. Army Engineer Waterways Experiment Station.

Pelletier, J.P. 1995. Demonstrations and commercial applications of innovative sediment removal technologies. Pp. 112–127 in Dredging, Remediation, and Containment of Contaminated Sediments. ASTM STP 1293. Philadelphia: American Society of Testing and Materials.

Pilarczyk, K.W. 1994. Novel Systems in Coastal Engineering, Geotextile System and Other Methods: An Overview. Rijkswaterstaat, Road and Hydraulic Engineering Division, Delft, The Netherlands.

Pritchard, P.H., and C.F. Costa. 1991. EPA's Alaska oil spill bioremediation project. Environmental Science and Technology 25(3):372–379.

Rubin, Debra K. 1995. Estimating: Cleanup costing seeks order. Engineering News-Record 235(13):46.

Sedlak, D.L., and A.W. Andren. 1994. The effect of sorption on the oxidation of polychlorinated biphenyls (PCBs) by hydroxyl radical. Water Research 28(5):1207–1215.

Shields, F., and R. Montgomery. 1984. Fundamentals of capping contaminated dredged material. Pp. 446–460 in Dredging '94: Proceedings of the 2nd International Conference on Dredging and Material Placement. E.C. McNair, Jr., ed. New York: American Society of Civil Engineers.

Stern, E., J. Olha, A.A. Massa, and B. Wisemiller. 1994. Recent assessment and decontamination studies of contaminated sediments in the New York/New Jersey Harbor. Pp. 458–467 in Dredging '94: Proceedings of the 2nd International Conference on Dredging and Material Placement. E.C. McNair, Jr., ed. New York: American Society of Civil Engineers.

Sturgis, T., and D. Gunnison. 1988. A Procedure for Determining the Cap Thickness for Capping Subaqueous Dredged Material Deposits. Technical Note EEDP-0109. Vicksburg, Mississippi: U.S. Army Engineer Waterways Experiment Station.

Sukol, R.B., and G.D. McNelly. 1990. Workshop on Innovative Technologies for Treatment of Contaminated Sediments: Summary Report. EPA-600/2-90-054. Risk Reduction Engineering Laboratory, Office of Research and Development. Cincinnati: U.S. Environmental Protection Agency.

Sumeri, A. 1984. Capped in-water disposal of contaminated dredged material. Pp. 644–653 in Dredging '84: Proceedings of the 1st International Conference on Dredging and Material Disposal. R.L. Montgomery and J.W. Leach, eds. New York: American Society of Civil Engineers.

Taylor, A. 1995. Bayou Bonfouca Superfund cleanup project mission completed. World Dredging, Mining, and Construction 31(8):16–17.

Tetra Tech, Inc., and D. Averett. 1994. Options for Treatment and Disposal of Contaminated Sediments from New York/New Jersey Harbor. Miscellaneous Paper EL-94-1. Vicksburg, Mississippi: U.S. Army Engineer Waterways Experiment Station.

Thibodeaux, L.J. 1989. Theoretical Models for Evaluation of Volatile Emissions to Air During Dredged Material Disposal with Application to New Bedford Harbor, Massachusetts. Miscellaneous Paper EL-89-3. Vicksburg, Mississippi: U.S. Army Engineer Waterways Experiment Station.

Thibodeaux, L.J., D.D. Reible, W. Bosworth, and L. Sarapas. 1990. A Theoretical Evaluation of the Effectiveness of Capping PCB Contaminated New Bedford Harbor Sediment. Hazardous Waste Research Center. Baton Rouge: Louisiana State University.

Thoma, G. 1994. Summary of the Workshop on Contaminated Sediment Handling, Treatment Technologies, and Associated Costs held April 21–22, 1994. Background paper prepared for the Committee on Contaminated Sediments, Marine Board, National Research Council, Washington, D.C.

Toronto Harbor Commission. 1993. Report on the Treatment of the Toronto Harbour Sediments at the THC Soil Recycling Plant. Toronto, Ontario, Canada: Wastewater Technology Centre.

Truitt, C.L. 1986. The Duwamish Waterway Capping Demonstration Project: Engineering Analysis and Results of Physical Monitoring. Technical Report D-86-2. Long-Term Effects of Dredging Operations Program. Vicksburg, Mississippi: U.S. Army Engineer Waterways Experiment Station.

Truitt, C.L. 1987a. Engineering Considerations for Capping Subaqueous Dredged Material Deposits—Background and Preliminary Planning. Environmental Effects of Dredging. Technical Note EEDP-01-3. Vicksburg, Mississippi: U.S. Army Engineer Waterways Experiment Station.

Truitt, C.L. 1987b. Engineering Considerations for Capping Subaqueous Dredged Material Deposits—Design Concepts and Placement Techniques. Environmental Effects of Dredging. Technical Note EEDP-01-4. Vicksburg, Mississippi: U.S. Army Engineer Waterways Experiment Station.

U.S. Army Corps of Engineers (USACE). 1987. Confined Disposal of Dredged Material—Engineering Manual. EM-1110-2-5027. Washington, D.C.: USACE.

USACE. 1993. Dredging and Dredged Material Disposal. Engineer Manual. 1110-2-5025. Washington, D.C.: USACE.

USACE and EPA. 1992. Evaluating Environmental Effects of Dredged Material Management Alternatives—A Technical Framework. EPA 842-B-92-008. Washington, D.C.: USACE and EPA.

U.S. Department of the Army and USACE. 1995. Navstar Global Positioning System Surveying. Engineer Manual. EM 1110-1-1003. Washington, D.C.: USACE.

U.S. Army Engineer Buffalo District. 1993. Pilot-Scale Demonstrations of Thermal Desorption for the Treatment of Buffalo River Sediments, Assessment and Remediation of Contaminated Sediments (ARCS) Program. Great Lakes National Program Office. EPA 905-R93-005. Chicago: EPA.

U.S. Army Engineer Buffalo District. 1994. Pilot-Scale Demonstrations of Thermal Desorption for the Treatment of Ashtabula River Sediments, Assessment and Remediation of Contaminated Sediments (ARCS) Program. Great Lakes National Program Office. EPA 905-R94-021. Chicago: EPA.

U.S. Army Engineer Detroit District. 1994. Assessment and Remediation of Contaminated Sediments (ARCS) Program: Pilot-Scale Demonstration of Sediment Washing for the Treatment of Saginaw River Sediments. Great Lakes National Program Office. EPA 905-R94-019. Chicago: EPA.

Valent, P.J., and D.K. Young. 1995. Technical and Economic Assessment of Storage of Industrial Waste on Abyssal Plains. Paper Presented to the Marine Board, John C. Stennis Space Center, Mississippi, June 21–23, 1995.

van der Veen, R. 1995. Contaminated Sediment Remediation: Dredging Polluted Bed Materials: A Study of Environmentally Effective Dredging Methods. Directorate General for Public Works and Water Management (Rijkswaterstaat). North Sea Directorate, P.O. Box 5807, 2280 HV Rijswijk, The Netherlands. April.

van Oostrum, R.W. 1992. Dredging Contaminated Sediments in the Netherlands. Proceedings from the International Symposium on Environmental Dredging. Buffalo, New York: Erie County Environmental Education Institute, Inc.

Wardlaw, C. 1994. Interim results of Canada's Contaminated Sediment Treatment Technology Program. Background material provided for the Committee on Contaminated Marine Sediments. Workshop on Handling and Treatment Technologies and Associated Costs held April 21–22, 1994 in Chicago, Illinois.

Wenzel, J.G. 1994a. Feasibility and Availability of Equipment for Dredging Contaminated Sediments from the Palos Verdes Shelf and Slope. Saratoga, California: Marine Development Associates, Inc.

Wenzel J.G. 1994b. Dredging Equipment and Controls for Palos Verdes Shelf and Slope. MDA 93-001. Saratoga, California: Marine Development Associates, Inc.

West Harbor Operable Unit. 1992. Wyckoff/Eagle Harbor Superfund Site Record of Decision. Seattle, Washington: EPA, Region 10.

Wong, C.S., G. Sanders, D.R. Engstrom, D.T. Long, D.L. Swackhammer, and S.J. Eisenreich. 1995. Accumulation, inventory, and diagenesis of chlorinated hydrocarbons in Lake Ontario sediments. Environmental Science and Technology 29(10):2661–2672.

Zappi, M.E., and D.F. Hayes. 1991. Innovative Technologies for Dredging Contaminated Sediments. Miscellaneous Paper EL-91-20. Vicksburg, Mississippi: U.S. Army Engineer Waterways Experiment Station.

Zeman, A.J., S. Sills, J.E. Graham, and K.A. Klein. 1992. Subaqueous Capping of Contaminated Sediments: Annotated Bibliography. Burlington, Ontario, Canada: National Water Research Institute, Environment Canada.

6

Conclusions and Recommendations

The challenges to be overcome in the management of contaminated sediments are multifaceted, and there are no easy solutions. The problem is not intractable, however, as long as two key issues are addressed: forging partnerships to replace adversarial relationships; and changing laws, regulations, and practices.

To provide a framework for the committee's specific proposals, a number of general observations can be made based on the analysis presented in this report. Most important, there is no simple solution, although many people may assume there is, and there is no breakthrough technology on the immediate horizon for treating large volumes of contaminated sediments effectively and economically. Although in situ and handling technologies have been used with some degree of effectiveness, ex situ decontamination technologies are generally not affordable except when sediment volumes are small or when the benefits to public health or the environment are expected to be extremely high. Thus, near-term improvements in sediment management are likely to come from changes in the decision-making processes that will speed the implementation of solutions, improve the political acceptability of the management strategy and decisions, and apply systems engineering to reduce overall costs. In other words, there is no reason to delay urgent projects in anticipation of new technological solutions; decision makers should continue to try to make incremental improvements in the overall management process.

Some impediments to effective remedial action are legal and regulatory in nature. In some cases, the problems stem from how laws and regulations are interpreted rather than their original intent; but even these difficulties can impede the decision-making process. Some barriers could be removed through revisions to, or objectives-based application of, existing laws and regulations. Substantial uncertainties will remain, however, in methods for assessing the effects of contaminated sediments on human health and the environment and for evaluating the risks, costs, and benefits of various management options. In addition, intangible factors, such as social perceptions, will continue to have an important influence on the feasibility of particular options. The balanced consideration of risks, costs, and benefits can focus management decisions on a range of options, but qualitative judgment will continue to be the deciding factor.

The following formal conclusions and recommendations are organized into three broad areas where improvement is both necessary and possible: decision making, remediation technologies, and project implementation.

IMPROVING DECISION MAKING

Cost-effective management of contaminated sediments requires informed decisions about the levels of analysis and action required to characterize contaminated sites, to identify and manage appropriately the risks associated with sediment contaminants, and to confirm the results of remediation or containment through monitoring. Decision making is influenced by the statutory framework for remediating contaminated sediments, the interests of stakeholders, systems engineering considerations, and approaches to decision making. Improvements in each of these areas can contribute to better decision making.

Regulatory Constraints

Because the laws and regulations that affect the characterization, management, and monitoring of contaminated sediments were originally written to address other issues (such as water quality), contaminated sediments are often treated as an afterthought. As a result, barriers to sediment management may be imposed without technical justification. Moreover, regulations tend to emphasize the *mechanics* of tasks (e.g., placement location) rather than an appropriate balance of risks, costs, and benefits. These features of the regulatory framework can interfere with efforts to implement the best management practices and timely, cost-effective solutions.

For example, the three sets of regulations governing the evaluation of remedial alternatives use different approaches, and none fully considers either the degree of risk posed by contaminated marine sediments or the costs and benefits (i.e., economic and technical viability) of the various solutions. The MPRSA requires biological testing of dredged material to determine its inherent toxicity but does not fully consider site-specific considerations that may influence the exposure of organisms in the receiving environment, meaning that, at best, risk is considered only indirectly, and actual impacts are only approximated. This rigid approach may obstruct efforts to reach the best decision for a particular case and lead to the needless waste of scarce resources. The CWA procedures, which consider chemical and physical as well as biological characteristics in assessing whether the discharge of dredged material will cause unacceptable adverse impacts, are not risk-based, but at least they do not specify rigid pass-fail criteria; they are geared to the identification of the least environmentally damaging, practical (i.e., economically and technically viable) alternative. The Superfund remedial action program addresses risks and costs to some degree. An exposure assessment (but not a full risk analysis) is required to assess in-place risks, remedial alternatives are identified based on their capability to reduce exposure risks to an acceptable level, and the final selection involves choosing the most cost-effective solution. However, Superfund has no risk-based cleanup standards for underwater sediments.

Although inconsistencies among the three sets of regulations is not a major problem in and of itself, the lack of emphasis on risks, costs, and benefits impedes efforts to reach technically sound decisions about cost-effective management. One way to change the emphasis might be through legislation. For example, the U.S. Congress, by enacting and revising environmental laws as they apply to contaminated sediments, and the EPA and USACE, in implementing these laws, could adopt objectives-based approaches that reflect an appropriate balance among risks, costs, and benefits.

Conclusion. The evaluation of disposal and management options needs to be based on the fullest practical consideration of the relevant risk factors as well as on technological feasibility and economic viability.

Similar inattention to risk is evident in the permitting processes for sediment disposal. Currently, different types of permits must be secured for the placement of sediments in navigation channels or ocean waters as part of the construction of land or containment facilities (under the RHA), the dumping of sediment in the ocean (under the MPRSA), sediment disposal in inland waters or wetlands (under CWA), and the containment of contaminated sediments on land (under the RCRA). The regulations also distinguish between sediments removed during navigation dredging (CWA or MPRSA) and sediments excavated for environmental remediation (Superfund). In other words, the regulatory framework does not differentiate between the placement of contaminated sediments in an ecologically sensitive and commercially valuable shellfish bed and the deposition of contaminated sediments within the confining walls of an offshore containment dike or in the depths of an anoxic, deep ocean pit.

The committee can see little technical justification for the inconsistent regulation of contaminated sediments, given that neither the location of an aquatic disposal site (freshwater versus saltwater) nor the reason for the dredging (navigation versus environmental remediation) necessarily affects the risks posed by the in-place contamination. In the committee's view, the regulatory regime pays little or no attention to the question of risk, focusing instead on the types of activities to be carried out—removal, placement, or treatment. The problem has been eased in some cases by objectives-based interpretation of regulations, as demonstrated by the carefully considered solution in the Port of Tacoma case history.

Conclusion. The failure to regulate contaminated sediments based on systematic consideration of risk management is inefficient and leads to less than optimum expenditures of time and money.

Systematic, integrated decision making may also be undermined by regulations governing cost allocation and cost-benefit analysis. The federal government pays for a share of new-work dredging and all maintenance dredging through a user-fee mechanism but pays for none of the costs of sediment disposal. The local sponsors of federal navigation projects must bear the burden of identifying, constructing, operating, and maintaining the placement sites for dredged material, under the project cooperation requirement of the WRDA of 1986. This inconsistent approach to cost sharing may foster irrational allocations of scarce resources. Because the project sponsor must pay for disposal on land, whereas open-water disposal is paid for by the federal government as a component of dredging costs, the WRDA provision creates a strong preference for the latter, regardless of whether it is in society's (or the environment's) best interest. Furthermore, a local sponsor bearing the full burden of disposal costs has little incentive to seek out opportunities for the beneficial uses of dredged materials, which usually add to the project cost and may benefit third parties, such as the public. Additional inconsistencies are introduced in the area of cost-benefit analysis. Currently, an elaborate weighing of costs and benefits must be performed for new-work dredging. But no similar cost-benefit analysis is required for either maintenance dredging or the placement of dredged material.

Conclusion. The cost effectiveness of managing dredged material would be improved if the various elements of federal projects—including dredging and placement—were subject to consistent approaches to cost-sharing and to cost-benefit analysis.

One option Congress might consider is amending the project cooperation requirement of WRDA as it relates to financial responsibility for the construction of land-based or aquatic sediment containment facilities, so that consistent cost-sharing formulas apply to dredging and placement for federal projects. To ensure that costs are controlled, dredging and disposal for a project could both be subjected to cost-benefit analysis (preferably on a combined basis) and to the application of a systems engineering approach.

Outreach to Stakeholders and Consensus Building

To be successful, a remediation project needs strong proponents, whether federal agencies or ports. The identification and timely implementation of effective solutions also depend heavily on how project proponents interact with stakeholders. Many parties—including government agencies at all levels, environmental groups, and members of the local community—have interests or stakes in the management of contaminated sediments, but they may have different perspectives on the problem and proposed solutions. Because any participant in the decision-making process can block or delay remedial action, project proponents need to identify all stakeholders and build consensus among them. The development of consensus can be fostered by using various tools, including mediation, negotiated rule making, collaborative problem solving, and effective communication of risks.

Conclusion. It is impossible to legislate agreement on issues that are inherently subject to debate. Therefore, the early involvement of stakeholders is important for heading off disagreements and for building consensus. Project proponents need to identify all stakeholders early in the decision-making process and continue to devote significant efforts to building relationships with stakeholders and reaching consensus.

Systems Engineering

The complexity of decision making can be accommodated by a systems approach in which interrelated issues and tasks are considered in concert. Systems engineering and analysis are widely used but have seldom been applied rigorously to decisions about the management of contaminated sediments. The overall goal is to manage the system in such a way that the results are optimized. In particular, a systems approach is advisable for the selection and optimization of interim and long-term control technologies. Limited resources and the high cost of technology demand that trade-offs be made and that remediation solutions be optimized.

Conclusion. Systems engineering techniques can enhance the cost-effectiveness of the management of contaminated sediments. The use of systems engineering in choosing a remediation technology will help ensure that the solution meets all removal, containment, transport, and placement requirements while satisfying environmental, social, and legal requirements.

Approaches to Decision Making

Three approaches can be applied to inform and improve decision making about contaminated sediments, particularly with respect to weighing the risks,

CONCLUSIONS AND RECOMMENDATIONS

costs, and benefits of proposed solutions: risk analysis, cost-benefit analysis, and decision analysis. The application of these approaches requires time and training.

Risk Analysis

To ensure the cost-effective management of risks to human health and the environment, risk analysis (the combination of risk assessment and risk management) needs to be used throughout the management process. Currently, risk analysis is not fully applied in the context of managing contaminated sediments. Typically, risks are assessed only at the beginning of the decision-making process, with the focus on in-place contamination. Risks are seldom reassessed after the implementation of solutions. As a result, capabilities for evaluating management strategies and remediation technologies are limited. Extended application of risk analysis, particularly in the selection and evaluation of management strategies, would not only inform decision making in specific situations but would also provide data that could be used to evaluate generic approaches and plan future projects. The results of risk analysis are also essential ingredients for other important decision-making tools, such as cost-benefit analysis and decision analysis.

Conclusion. The committee recognizes that there are uncertainties but believes that risk analysis techniques can be applied more widely than they are now in the sediment management process to improve decision making, particularly with respect to the selection and evaluation of management strategies and remediation technologies.

The scientific underpinning of risk analysis as applied to contaminated sediments also requires attention. A fundamental uncertainty in current approaches lies in the methods used to assess initial risks. The effects-based testing methods currently used are being improved to include protocols for both acute and chronic effects as a basis for making decisions concerning the placement of dredged materials. The USACE and EPA are also moving toward applying formal risk assessment to the results of bioaccumulation tests. Risk assessment will provide improved end-points, but there will still be a need to understand and interpret biological end-points in a regulatory context to determine unacceptable adverse effects.

Conclusion. Continued development and risk-based interpretation of the results of effects-based testing methods would promote cost-effective management, support the quantitative evaluation of the performance of remediation technologies, and assist in the assessment and selection of options for sediment disposal and options for the beneficial use of treated sediments.

Cost-Benefit Analysis

Cost-benefit analysis is not applied widely to the management of contaminated sediments. It is currently used only for major new navigation dredging projects and is usually narrow in scope. However, cost-benefit analysis could be used in many cases to help identify the best strategy for managing contaminated sediments. From an economic standpoint, the best strategy is one in which benefits outweigh the costs by as the much as possible. The costs involved in the management of contaminated sediments are difficult to calculate and cannot be measured precisely, but a comprehensive cost-benefit analysis may be worth the effort in very expensive or extensive projects. Informal estimates or cost-effectiveness analysis may suffice for smaller projects. Current federal guidelines for the computation and use of benefit and cost data (generally confined to the navigation dredging context) are neither comprehensive nor applied systematically to the management of contaminated sediments. For example, the guidelines do not take into account the economic effects of shifts in transportation patterns or changes in the prices of navigation services.

Conclusion. More extensive use of appropriate methods for cost-benefit analysis have the potential to improve decision making.

Decision Analysis

Methods are needed for balancing consideration of the risks, costs, and benefits of various sediment management strategies. One tool that can help resolve problems with multiple variables is decision analysis, which uses both factual and subjective information to evaluate the relative merits of alternative courses of action. This technique could be particularly valuable in certain situations because it can accommodate more variables (including uncertainty) and different perspectives than techniques like cost-benefit analysis that measure single outcomes. Decision analysis can also be a consensus-building tool because it enables stakeholders to explore subjective elements of contaminated sediments problems and perhaps find common ground. However, because it is technical in design and involves complex, logical computations, decision analysis is probably worth the effort only in highly contentious situations in which stakeholders are willing to devote enough time to gain confidence in the approach.

Conclusion. Decision analysis could be used to help balance consideration of the risks, costs, and benefits of various management strategies in situations in which the issues are exceptionally complex and divisive.

Recommendations for Improving Decision Making

In addition to the suggestions for statutory changes that have already been made, the committee makes the following recommendations.

Recommendation. The Environmental Protection Agency and the U.S. Army Corps of Engineers should continue to develop uniform or parallel procedures to address the environmental and human health risks associated with the freshwater, marine, and land-based disposal, containment, or beneficial reuse of contaminated sediments.

Recommendation. Because consensus building is essential for project success, federal, state, and local agencies should work together with appropriate private-sector stakeholders to interpret statutes, policies, and regulations in a constructive manner so that negotiations can move forward and sound solutions are not blocked or obstructed.

Recommendation. To facilitate the application of decision-making tools, the Environmental Protection Agency and U.S. Army Corps of Engineers should: (1) develop and disseminate information to stakeholders concerning the available tools; (2) use appropriate risk analysis techniques throughout the management process, including the selection and evaluation of remediation strategies; and (3) demonstrate the appropriate use of decision analysis in an actual contaminated sediments case.

Recommendation. The USACE should modify the cost-benefit analysis guidelines and practices it uses to ensure the comprehensive, uniform treatment of issues involved in the management of contaminated sediments.

IMPROVING REMEDIATION TECHNOLOGIES

Technologies for remediating contaminated sediments are at various stages of development. Sediment handling technologies are the most advanced, although there are benefits to be realized by improvements in the precision of dredging (and, concurrently, in site characterization). The state of practice for in situ controls ranges from immature (e.g., bioremediation) to rapidly evolving (e.g., capping). A number of ex situ treatment technologies exist that probably can be applied successfully to contaminated sediments, but additional R&D and full-scale demonstrations are needed to determine their effectiveness. Moreover, these technologies are expensive, and it is not certain whether unit costs would drop significantly in full-scale implementation. Ex situ containment, however, is commonplace.

Overall, the uneven state of the art suggests that technologies need to be selected, combined, and optimized using a systems engineering approach. Although 100 percent effectiveness is not possible, available technologies do offer adequate solutions.

A key factor in determining the utility of a remediation technology is cost. Therefore, cost issues are addressed before specific technologies.

Engineering Costs of Cleanup

The engineering costs of cleanup depend not only on the type of approach used but also on the number of steps involved—the more handling, the higher the cost. The costs of removing and transporting contaminated sediments (generally less than $15 to $20/yd^3) tend to be higher than the costs of conventional navigation dredging for "clean" sediments (seldom more than $5/yd^3) but much lower than the costs of treatment (often more than $100/yd^3). Reducing volume (i.e., removing only sediments that require treatment and entraining as little water as possible) offers greater cost savings than increasing production rates. Improved site characterization coupled with precision dredging techniques is particularly promising for reducing volume. Treatment costs may be reduced through pretreatment. For example, silt- and clay-sized particles may be separated from cleaner sand using hydrocyclones. However, the cost savings vary depending on the proportion of fine-grained sediments requiring further treatment, as well as the cost of that treatment. Pretreatment is usually worthwhile only when the sediment contains a substantial fraction of relatively clean sand.

Although post-cleanup data on actual costs are limited because of the small number of completed projects, numerous cost projections are available for approved or proposed projects. These figures, in the judgment of the committee, are sufficient for an evaluation of the practicality of various technologies.

Conclusion. Many contaminated sediments can be managed effectively using natural recovery, capping, or containment. Where remediation is necessary, high-volume, low-cost technologies are the first choice, if they are feasible. Because treatment is expensive, reducing volume is very important. At the current state of practice, treatment is usually justified only for relatively small volumes of highly contaminated sediments, unless there are compelling public health or natural resource considerations. Advanced treatment processes are too costly in the majority of cases of (typically low-level) contamination. The unit cost of advanced treatments will probably decline slightly as these technologies move through the demonstration phase, but it is unlikely to become competitive with the cost of less-expensive technologies, such as containment.

Problems with available cost data include the lack of standardized documentation and the lack of a common basis for defining all relevant benefits and costs.

The data are inconsistent with respect to the types of costs included and the units of measure (e.g., cubic yards, tons, hectares), and geographical variations in cost are not taken into account. The problem stems in part from the lack of a formal structure for reporting cost data. Even if good cost data were available, measures of effectiveness must be improved before reliable comparative analyses of technologies can be made.

Conclusion. Improved cost information is needed for full-scale remediation systems for fair, overall comparisons and to provide benchmarks for R&D and systems design. Although the lack of reliable cost data does not preclude project planning, better cost information would contribute to sound decision making.

Remediation Technology Options

In Situ Controls

In situ management offers the potential advantage of avoiding the costs and material losses associated with the excavation and relocation of sediments. Among the inherent disadvantages of in situ management is that it is seldom feasible in navigation channels that are subject to routine maintenance dredging. Another limitation is that monitoring needs to be an integral part of any in situ approach to ensure effectiveness over the long term.

Natural recovery is a viable alternative under some circumstances. It offers the advantages of low cost and, in certain situations, the lowest risk of human and ecosystem exposure to sediment contamination. Natural recovery is most likely to be effective where surficial concentrations of contaminants are low, where surface contamination is being covered over rapidly by cleaner sediments, or where other processes destroy or modify the contaminants thus decreasing contaminant releases to the environment over time. For natural recovery to be relied upon with confidence, the physical, chemical, and hydrological processes at a site need to be characterized adequately (although chemical movements cannot be quantified completely). Extensive site-specific studies may be required for this.

Conclusion. For many projects, natural recovery is a viable option. It may be the optimum solution where surficial concentrations of contaminants are low, where surface contamination is being covered over rapidly by cleaner sediments, or where contaminated sediment is modified by natural chemical or biological processes and the release of contaminants to the environment decreases over time. A better understanding of natural processes is needed, and models need to be verified through long-term monitoring.

The advantages of in situ capping are that it isolates the contaminants and may protect against sediment resuspension. At appropriate locations, capping materials can be emplaced readily and, if necessary, repaired. In situ capping requires that the original bed be able to support the cap, that suitable capping materials to create the cap are available, and that hydraulic conditions (including water depth) permit cap emplacement and will not compromise the integrity of the cap. Changes in the local substrate, benthic community structure, or bathymetry at a depositional site may subject the cap to erosion. These changes, among others, need to be verified by short-term pre-project and long-term post-project monitoring. A regulatory barrier to the use of capping is the language of Superfund legislation (§121[b]), which gives preference to "permanent" controls. Capping is not considered by regulators to be a permanent control, but the available evidence suggests that properly managed caps can be effective in reducing risks associated with underwater sediments. Furthermore, capping may be preferable to some other strategies because it is relatively inexpensive and easy to implement, and it capitalizes on the tendency of contaminants to remain bound to sediment particles and to settle in low-energy sinks.

Conclusion. When natural recovery is not feasible, capping may be an appropriate way to reduce bioavailability by minimizing contaminant contact with the benthic community. The efficacy of capping needs to be monitored, not only to ensure that risks are reduced, but also to gather data that can be used to advance the state of practice. The appropriate use of capping might be advanced if it were viewed as a permanent solution in the Superfund context.

In situ immobilization and the chemical treatment of contaminated sediments have not been demonstrated successfully in the marine environment, although the concept is attractive because the cost of sediment removal would be avoided. In situ chemical treatment would be complicated by the need to isolate sediments from the water column during treatment, by inaccuracies in reagent placement, and by the need for long-term follow-up monitoring. Other constituents in this sediment (e.g., natural organic matter, oil and grease, metal sulfide precipitates) could interfere with chemical oxidation. Immobilization techniques may not be applicable to fine-grained sediments with a high water content.

Conclusion. Although there are conceptual advantages to in situ chemical treatment, considerable R&D will be needed before successful application can be demonstrated.

Biodegradation has been observed in soils, in groundwater, and along shorelines contaminated by a variety of organic compounds (e.g., petroleum products, benzene, toluene and xylene, PCBs, polyaromatic hydrocarbons, chlorinated phenolics, pesticides). However, biodegradation in subaqueous environments presents a number of significant microbial, geochemical, and hydrological problems and has yet to be demonstrated.

Conclusion. Using bioremediation to treat in-place marine sediments, although theoretically possible, requires further R&D because it raises a number of significant microbial, geochemical, and hydrological issues that have yet to be resolved.

Sediment Removal Technologies

Efficient hydraulic and mechanical methods are available for the removal and transport of sediments for ex situ remediation or confinement. Most dredging technologies that can be used to remove contaminated sediments have been designed for large-volume navigation dredging rather than for the precise removal of hot spots. Promising technologies for precision control include electronically positioned dredge heads and bottom-crawling hydraulic dredges. The latter also may offer the capability of dredging in depths beyond the standard maximum operating capacity. The cost effectiveness of dredging innovations can best be judged through side-by-side comparisons to current technologies.

Conclusion. Because of the high cost of ex situ treatment relative to dredging, dredges need to be made widely available that can remove sediments at near in situ densities and that have the capability for the precise removal of contaminated sediments, so that the capture of clean sediments and water can be limited, thus reducing the volume of dredged material requiring containment or treatment.

Ex Situ Technologies

Containment technologies, particularly CDFs, have been used successfully in numerous projects. A CDF can be effective for long-term disposal if it is well designed to contain sediment particles and contaminants and if a suitable site can be found. A CDF can also be a treatment or interim storage facility where sediments can be separated for varying levels of treatment and, in some cases, for beneficial reuse. Costs of CDFs are reasonable; in some parts of the country reusing CDFs may be cheaper than building new ones. Under some circumstances, CDFs can foster development in urban areas. Disadvantages of this technology include the imperfect methods of controlling pathways of contaminant release. Improved long-term monitoring methods are also needed.

Conclusion. Research is needed to improve the control of contaminant releases, to improve long-term monitoring methods, and to improve techniques for preserving the capacity of existing CDFs.

Contained aquatic disposal (CAD) is an appropriate method for managing contaminated sites in shallow waters where in situ capping is not possible and for containing moderately contaminated material from navigation dredging. Some advantages of CAD are that it can be performed with conventional dredging equipment and that the chemical environment surrounding the cap is not changed. The committee could not locate any useful estimates of actual costs. A disadvantage is the possible loss of small quantities of contaminated sediments during placement operations. Improved tools are needed for designing sediment caps and armor layers and for evaluating long-term stability and effectiveness.

Conclusion. Construction of CADs on or near contaminated sites is likely to be acceptable, but the applications have not been explored fully. Research is needed to improve design tools and long-term monitoring methods and to control contaminant losses and determine their effects and associated risks.

Scores of ex situ treatment technologies have been bench and pilot tested, and some may warrant broader testing in marine systems, depending on their applicability to particular problems. Ex situ treatment in general is more promising than the same treatment in situ because conditions can be controlled more effectively in a contained facility. Chemical separation, thermal desorption, and immobilization technologies have been used successfully but are expensive, complicated, and limited to treating certain types of sediments. Because of extraordinarily high unit costs, thermal and chemical destruction techniques do not appear to be cost-effective near-term approaches for remediating large volumes of contaminated dredged sediment.

Conclusion. R&D on ex situ treatment technologies is warranted in the search for reasonable possibilities for the cost-effective treatment of large volumes of sediment. Bench and pilot testing of ex situ treatment technologies, and eventually full-scale demonstrations in marine systems, are needed to improve cost estimates, resolve technical problems, and improve treatment effectiveness.

Ex situ bioremediation, which is not as far along in development as other ex situ treatments, presents an enormous number of technical problems, making it a costly option, at least initially, for treating contaminated sediments. However, if the technical problems can be resolved, ex situ bioremediation has the potential, over the long term, to provide cost-effective remediation of large volumes of sediment. Ex situ bioremediation has greater potential than the in situ approach because conditions can be controlled better in a contained facility. The technique has been demonstrated on a pilot scale with some success, but complex questions about how to engineer the system remain.

Conclusion. In the search to develop reasonable, cost-effective treatments for large volumes of sediment, R&D on ex situ bioremediation is warranted. The initial focus should be on developing and verifying methods for marine systems.

Remediation Technology Research, Development, Testing, and Demonstration

It is unrealistic to expect dramatic breakthroughs that would substantially reduce the cost of large-scale ex situ treatment. In the near term, therefore, the optimal use of existing technologies—optimizing dredging and containment technologies through systems engineering, for example—appears to be the best way to enhance management effectiveness. Yet continuing efforts to identify, develop, and demonstrate new and improved remediation approaches are crucial to improving the management of contaminated sediments. Major technological challenges to be overcome include high costs, inadequate methods of predicting effectiveness, and the lack of extensive testing of many advanced treatment technologies, as well as the uncertainty associated with any innovative approach.

The value of demonstration projects and of peer review of proposed technologies was confirmed by the Assessment and Remediation of Contaminated Sediments research and planning program in the Great Lakes. The development and use of innovative technologies might be promoted through side-by-side demonstrations with current technologies, an approach used in the EPA's Superfund Innovative Technology Evaluations program, which evaluates cleanup technologies for toxic and hazardous waste.

Conclusion. Additional R&D and demonstration projects are needed to improve existing remediation technologies and to reduce the risks associated with the development and use of innovative approaches to treating marine sediments. The development and wide use of cost-effective, innovative solutions would be advanced by (1) the peer review of proposals for R&D on new technologies for handling, containing, and remediating sediments and (2) the establishment of mechanisms for side-by-side demonstrations of new and current technologies.

Recommendations for Improving Long-Term Controls and Technologies

Based on the conclusions regarding the engineering costs of cleanup, in situ controls, sediment removal technologies, ex situ technologies, and remediation technology R&D, the committee offers the following recommendations, presented in order of priority:

Recommendation. The Environmental Protection Agency and U.S. Army Corps of Engineers should develop a program to support research and development and to demonstrate innovative technologies specifically focused on the placement, treatment, and dredging of contaminated marine sediments. Innovative technologies should be demonstrated side by side with the current state-of-the-art technologies to ensure direct comparisons. The results of this program should be published in peer-reviewed publications so the effectiveness, feasibility, practicality, and cost of various technologies can be evaluated independently. The program should span the full range of research and development, from the concept stage to field implementation.

Recommendation. The U.S. Army Corps of Engineers and Environmental Protection Agency should develop guidelines for calculating the costs of remediation systems, including technologies and management methods, and should maintain data on the costs of systems that have actually been used. The objective should be to collect and maintain data for making fair comparisons of remediation technologies and management methods based on relative costs as well as their effectiveness in reducing risks to human health and ecosystems.

Recommendation. The Environmental Protection Agency and U.S. Army Corps of Engineers should support research and development to reduce contaminant losses from confined disposal facilities and confined aquatic disposal, to promote the reuse of existing confined disposal facilities, and to improve tools for the design of confined disposal facilities and confined aquatic disposal systems and for the evaluation of long-term stability and effectiveness.

Recommendation. The Environmental Protection Agency and U.S. Army Corps of Engineers should sponsor research to develop quantitative relationships between the availability of contaminants and the corresponding risks to humans and ecosystems. The overall goal should be to enable project evaluation using performance-based standards, specifically the risk reduction from in-place sediments; disturbed sediments; capped sediments; confined disposal facilities and confined aquatic disposal; and sediments released following physical, chemical, thermal, and biological treatments.

Recommendation. The Environmental Protection Agency and U.S. Army Corps of Engineers should support the development of monitoring tools to assess the long-term performance of technologies that involve leaving contaminants in or near aquatic environments. Monitoring programs should be demonstrated with the goal of ensuring that risks have been reduced through contaminant isolation.

IMPROVING PROJECT IMPLEMENTATION

Improvements in decision making and remediation technologies would go a long way toward ensuring the cost-effective management of contaminated sediments, but assorted practical issues also need to be addressed to remove constraints on project implementation. First, responsibilities for source control need to be allocated fairly. Second, improved site assessment capabilities need to be developed and implemented to enhance overall cost effectiveness. Third, appropriate interim controls need to be used to reduce high risks until long-term solutions have been found. Fourth, incentives are needed to encourage the beneficial reuse of contaminated sediments; this would promote public acceptance, cost effectiveness, and address the problem of the shortage of disposal space.

Responsibility for Source Control

By commercial necessity, ports are located in quiescent waters, which are also natural sediment traps. Because accumulations of sediment interfere with deep-draft navigation, ports need to dredge periodically. If the sediments to be dredged are contaminated, then ports are responsible for sediment placement and any necessary remediation. The suggested revisions in cost-sharing formulas for dredging and placement projects would relieve some of the burden on ports but would not address the issue of source control. Upstream generators of contaminants often cannot be identified or held accountable, leaving ports to manage the problem.

States, which benefit economically from dredging and which customarily engage in watershed management, could assume some of the responsibility for source control if they are not already in charge of ports. Under the CWA (§303), the EPA and the states set total maximum daily loads for waterway segments and develop local allocations for sources of pollution in an effort to control water pollution. This approach, although it might be difficult to implement, could be expanded to address sources of sediment contamination. To foster such an approach, congressional initiatives (i.e., CWA reauthorization legislation) requiring watershed planning and management to control sources of water pollution could take into consideration upstream contributions to downstream sediment. In addition, government regulators and ports could explore all available legal and enforcement tools for forcing polluters to bear a fair share of cleanup costs.

Conclusion. Ports bear an unfair share of the responsibilities for the remediation and placement of contaminated sediments; project implementation could be facilitated by transferring the burden for source control to states (where applicable) and polluters.

Site Characterization Needs and Technologies

Site assessments need to be comprehensive and accurate enough to define contamination chemically and geographically. Inaccurate or incomplete assessments can leave areas of unidentified contamination that continue to pose unmanaged risks. Another compelling argument for accurate site assessment is the need to control remediation costs; precise site definition is necessary to facilitate the removal of only those sediments that are contaminated, thus reducing the volume of sediment that requires expensive remediation. But site assessment is also expensive. The challenge lies in selecting the technique(s) and level of detail appropriate for the management phase and site in question.

In a systems approach to sediment management, overall cost effectiveness is maximized when the accuracy of the site assessment is matched to the precision of the dredging equipment. However, the high costs of physical coring and sample testing (the most common site characterization approach) hinders precise definition of either horizontal or vertical contaminant distributions, often leading to the removal and "remediation" of large quantities of "uncontaminated" sediments at unnecessarily high costs.

Conclusion. New and improved techniques are needed to reduce the costs and improve the precision of site assessment.

Acoustic profiling helps define the thickness and distribution of disparate sediment types. Because contaminants tend to be associated with fine-grained material, acoustic profiling may provide for the cost-effective, remote surveying of contaminated sediments, thereby increasing the precision and accuracy of site assessment. Additional R&D is needed, however. Meanwhile, chemical sensors used for soil and groundwater site assessments are being adapted for marine use. Examples include X-ray fluorescence for the detection of metals in sediments and fiber-optic chemical sensors.

Conclusion. Remote-sensing technologies, including rapid and accurate sensors, could reduce the costs and improve the precision of site assessments.

Interim Controls

The use of interim controls may be advisable when sediment contamination poses an imminent danger and an immediate risk reduction is required. More complete remediation solutions usually require considerable time to implement (3 to 15 years according to the committee's case histories). Identifying an imminent hazard is usually a matter of judgment, but in general an imminent hazard exists when contamination levels exceed a threshold level by a significant amount.

Both administrative interim controls (e.g., signs, health advisories) and structural controls have been used, and additional structural controls, such as CDFs for temporary storage, appear promising. However, few data are available concerning the effectiveness of interim controls because few have been used, and even fewer have been evaluated.

Conclusion. Although few data are available concerning the effectiveness of interim controls, a number of measures appear to be practical and are likely to reduce risk to some (albeit unknown) degree. Other advantages include low cost and ease of implementation.

Promotion of Beneficial Uses

Dredged material has been used for many beneficial purposes, including the creation of islands for seabird nesting, landfills for urban development, and wetlands, as well as for beach nourishment and shoreline stabilization. The policy focus and most of the experience has been with clean sediments, but the beneficial reuse of contaminated sediments is both possible and worthwhile. Reuse can provide alternatives to increasingly scarce disposal sites while making management plans more attractive, or at least palatable, to stakeholders. Some contaminated sites have been successfully transformed into wetlands, and research is under way on the safe use of contaminated sediments for various purposes, including "manufactured" topsoil and landfill covers. However, funding for this type of research is limited, and technical guidelines have yet to be developed. Other barriers include the USACE policy of selecting lowest-cost disposal options with little regard for the possibilities of beneficial use and disputes over whether the incremental costs of beneficial use should be borne by the project proponent or the beneficiary.

Conclusion. The beneficial use of dredged contaminated material, although constrained by both the contamination and the poor structural properties of most contaminated sediments, can provide much-needed disposal alternatives and enhance the social acceptance of a project.

Recommendations for Improving Project Implementation

Based on the conclusions regarding source control, site characterization, interim controls, and the beneficial use of contaminated sediments, the committee offers the following recommendations, presented in order of priority:

Recommendation. The U.S. Army Corps of Engineers should revise its policies to allow for the implementation of placement strategies that involve the beneficial use of contaminated sediments even if they are not lowest cost alternatives. In addition, regulatory agencies involved in contaminated sediments disposal should develop incentives for and encourage implementation of beneficial use alternatives.

Recommendation. Funding should be continued for research and development of innovative beneficial uses for contaminated sediments and the development of technical guidelines and procedures for environmentally acceptable, beneficial reuse.

Recommendation. Federal and state regulators, as well as ports, should investigate the use of appropriate legal and enforcement tools to require upstream contributors to sediment contamination to bear a fair share of cleanup costs.

Recommendation. The Environmental Protection Agency and U.S. Army Corps of Engineers should conduct joint research and develoment projects to advance the state of the art in site assessment technologies. Objectives should include the identification and development of advanced survey approaches and new and improved chemical sensors for both surveying and monitoring.

Recommendation. The U.S. Army Corps of Engineers should support demonstrations of innovative site assessment technologies. Remote sensing technologies should be demonstrated in an integrated survey operation at a major contaminated sediment site. The project should demonstrate the capability of accurately defining a hot spot or larger critical area that requires either in situ treatment or accurate removal for ex situ treatment or placement.

APPENDICES

APPENDIX A

Biographical Sketches of Committee Members

Henry J. Bokuniewicz, co-chair, is a professor at the Marine Sciences Research Center of the State University of New York at Stony Brook. Dr. Bokuniewicz has authored or co-authored numerous papers on sediment transport and deposition, sediment mass balance, and the effects of storm and tidal energy. His current research focuses on the effects of resuspension on containment availability for dredged material, benthic studies associated with containment, the prediction of tidal circulation and hydrodynamics, and criteria for the selection of placement sites for dredged material. Dr. Bokuniewicz also served on a previous National Research Council committee that addressed contaminated sediments issues. He received a B.A. degree from the University of Illinois and M. Phil. and Ph.D. degrees from Yale University.

Kenneth S. Kamlet, co-chair, is a lawyer with Linowes and Blocher and an environmental scientist with more than 20 years of experience in environmental toxicology and regulation. Mr. Kamlet has been a principal and member of A.T. Kearney Inc.'s environmental, health, and safety practice and director of the National Wildlife Federation's Pollution and Toxic Substances Division. He has devoted much of his career to public policy and technical issues surrounding ocean dumping and the navigation dredging and remediation of sediments contaminated with toxicants, and he has published numerous papers on these and related topics. Mr. Kamlet chaired a previous National Research Council study on contaminated sediments. Mr. Kamlet has a B.S. degree in biology from the City College of New York, an M. Phil. degree in biology from Yale University, and a J.D. degree from the University of Pennsylvania.

W. Frank Bohlen is a professor of physical oceanography in the Department of Marine Sciences at the University of Connecticut, Groton. Dr. Bohlen is an expert on turbulence and sediment transport processes and has authored several papers on sediment dispersion associated with the disposal of dredged material and the ocean dispersal of particulate wastes. He has served on many research and planning committees, including two National Research Council committees addressing marine particulate wastes and dredging. Dr. Bohlen has a B.S. degree from the University of Notre Dame and a Ph.D. degree from Massachusetts Institute of Technology and Woods Hole Oceanographic Institution.

J. Frederick Grassle, a marine ecologist and oceanographer, is director of the Institute of Marine and Coastal Sciences at Rutgers University. He is also associate director of the New Jersey Agricultural Experiment Station for Marine Science at Rutgers University. Previously he was a senior scientist at the Woods Hole Oceanographic Institution. His research focuses on population biology of marine benthic organisms and coastal, deep sea, and coral reef communities, and he has authored numerous articles and books on these topics. Dr. Grassle has served on numerous federal, state, and academic scientific advisory committees. He was an expert witness at a recent congressional hearing on the ocean dumping of dredged material, and, with the Institute of Marine and Coastal Sciences and the Port Authority of New York and New Jersey, he has convened conferences on the remediation of sediments. He is currently co-chair of the New York/New Jersey Harbor Estuary Program Scientific and Technological Advisory Committee. He also serves on the board of the New Jersey Marine Sciences Consortium and the editorial boards of the Marine Technology Society and the Society for Conservation Biology. Dr. Grassle is a fellow of the American Association for the Advancement of Science and the Explorer's Club. He received a B.S. degree from Yale University and a Ph.D. degree in zoology from Duke University.

Donald F. Hayes is an assistant professor of civil and environmental engineering at the University of Utah, Salt Lake City. Dr. Hayes' current research includes projects on predicting contaminant release during dredging, transport, and placement operations; wetlands design and operation for improving water quality; and modeling water quality changes in reservoirs. He has published numerous reports and papers related to managing contaminated sediments as well as the general management of dredged material. Dr. Hayes received B.S. and M.S. degrees in civil engineering from Mississippi State University and a Ph.D. degree in environmental engineering from Colorado State University.

James R. Hunt is an associate professor of environmental engineering at the University of California, Berkeley. Dr. Hunt has authored or co-authored numerous articles and papers on research addressing the sediment-water interface. His current research focuses on the transport and transformation of volatile organic solvents in unsaturated soils and flow-induced fluidization and resuspension of soft bottom sediments. Dr. Hunt serves on the Science Advisory Committee of the U.S. Environmental Protection Agency's Hazardous Substance Research Center for the Great Lakes and Mid-Atlantic regions, the Nonpoint Source Pollution Advisory Board for the Alexander Lindsay, Jr., Museum in California, and the National Water Research Institute's Research Advisory Board. Dr. Hunt has a B.S. degree in civil and environmental engineering from the University of California at Irvine, an M.S. degree in environmental engineering from Stanford University, and a Ph.D. degree in environmental engineering science from the California Institute of Technology.

Dwayne G. Lee is a principle project manager for The Parsons Infrastructure and Technology Group. Previously he was deputy executive director of development for the Port of Los Angeles, where he managed the engineering, construction management, environmental management, construction and maintenance, and 2020 program divisions. Prior to that he was a professor of engineering materials at the U.S. Military Academy, a congressional liaison and policy coordinator for the U.S. Army Corps of Engineers, and director of the U.S. Army Engineer Waterways Experiment Station. Mr. Lee received a B.S. degree from the U.S. Military Academy and an M.S. degree in space facilities engineering from the Air Force Institute of Technology.

Kenneth E. McConnell is a resource economist and professor in the Department of Agricultural and Resource Economics at the University of Maryland, College Park. His scholarly pursuits include economic assessments of natural resources and relationships between outdoor recreation benefits and the valuation of natural resources. His many publications include a forthcoming book on natural resource damage assessment, and he is on the editorial boards of various journals. He has been president of the Association of Environmental and Resource Economics, associate editor of the *Journal of Environmental Economics and Management and Leisure Sciences*, and a consultant to various federal and regional commissions. Dr. McConnell has a B.A. degree in economics from the University of Florida and a Ph.D. degree in economics from the University of Maryland.

Spyros P. Pavlou is technical director of environmental risk economics at URS Greiner, Inc. Until recently, he was director of Risk-Based Environmental Management, Ogden Environmental and Energy Services Corporation. He has more than 20 years of experience in environmental research, planning, and management; 7 years as a research professor at the Department of Oceanography, University of Washington; and 18 years as a professional consultant. His areas of specialty include environmental chemistry and the fate of contaminants in marine and freshwater environments; the development of sediment quality criteria and risk-based cleanup goals for hazardous waste site closures; cost-effectiveness analysis for the selection of remedial alternatives; and the application of integrated risk, cost-benefit, and decision analysis in environmental management. He has co-authored more than 30 papers, including peer-reviewed publications, conference proceedings, featured articles, and oral presentations. Dr. Pavlou received a B.S. degree in chemistry from the University of California at Los Angeles, an M.S. degree in physical chemistry from San Diego State University, and a Ph.D. degree in physical chemistry from the University of Washington.

Richard K. Peddicord is director of sediment management at EA Engineering, Science, and Technology Consulting Inc., an environmental consulting firm based in Hunt Valley, Maryland. Dr. Peddicord has 23 years of experience in all aspects of sediment management. His technical activities have included the evaluation and remediation of contaminated sediments, the assessment of potential impacts of dredging and disposal, regulatory review and negotiation, expert testimony, and participation in public meetings and hearings. Among his many credits, he compiled and edited the U.S. Environmental Protection Agency/U.S. Army Corps of Engineers joint manual for the evaluation of dredged material proposed for ocean dumping, and he led a human health and ecological risk assessment of dioxin associated with dredged material for the Port Authority of New York and New Jersey. He has served on many committees and boards addressing sediment toxicity and bioaccumulation. Dr. Peddicord received a B.S. degree in biology from Morehead State University and a Ph.D. degree in marine science from Virginia Institute of Marine Sciences.

Peter Shelley is the senior attorney and project director for Marine Resources and Water Resources of the Conservation Law Foundation, Inc., a public-interest conservation advocacy organization. His areas of concentration are water pollution and conservation, fisheries management, wetlands protection, pesticides, land-use management and planning, and marine resources. Mr. Shelley is a member of the Board of Directors and Policy Committee for Save the Harbor/Save the Bay, Inc., the Board of Directors of the Center for Coastal Studies, the Advisory Committee on Statewide Environmental Impact Report in Pesticide Use Rights-Of-Way, and the Massachusetts Coastwide Monitoring Project Steering Committee. He is a frequent lecturer, writer, and panelist on a range of environmental issues. Mr. Shelley received a B.A. degree from Hobart College and a J.D. degree from Suffolk University Law School.

Richard Sobel is vice president of technology and remediation for Clean Sites, Inc. He has 40 years of experience in the control and remediation of wastes in the air and water and on land. His expertise includes the technical, engineering, and operating aspects of chemical plants. Mr. Sobel's experiences at Clean Sites and at Allied Chemical Corporation, where he was director of environmental controls, have given him practical experience in the remediation of contaminated sites and an understanding of the industrial, government, and community points of view. He is affiliated with the American Institute of Chemical Engineers, was chairman of the Resource Conservation and Recovery Act (RCRA) Task Group of the Chemical Manufacturers Association (CMA), and was technical editor for CMA's *Chemecology*. He has published several papers on various aspects of RCRA and solid waste disposal practices in the chemical industry. Mr. Sobel has a B.Ch.E. degree from Cooper Union in New York and an M.Ch.E. degree from the University of Delaware.

Louis J. Thibodeaux is director of the U.S. Environmental Protection Agency's Hazardous Substance Research Center, South/Southwest, and the Jesse Coates Professor of Chemical Engineering at Louisiana State University in Baton Rouge. He has authored numerous papers and book chapters on the transport of contaminants from sediment beds and across the air-water interface. Dr. Thibodeaux is past chairman of the Environmental Division of the American Institute of Chemical Engineers. He is the author of a textbook entitled *Environmental Chemodynamics— Movement of Chemicals in the Air, Water, and Soil*, now in its second edition. He served recently on the National Research Council's Committee on Remedial Action Priorities for Hazardous Waste Sites. Dr. Thibodeaux has B.S., M.S., and Ph.D. degrees in chemical engineering from Louisiana State University.

James G. Wenzel, NAE, is president and chair of Marine Development Associates, Inc., a company he formed in 1984. Mr. Wenzel has 40 years of experience in the fields of ocean science, engineering, and development as an engineer, inventor, business executive, lecturer, and consultant. Formerly with Lockheed Corporation, he has been responsible for many ocean system and technology developments, including the Deep Quest research submarine, the U.S. Navy's deep submergence rescue vehicles, and the design and construction of deep ocean/ large object recovery systems. His environmental cleanup activities include the application of innovative technologies to the remediation of contaminated shelf sediments, corporate strategic planning, and ocean technology development. Mr. Wenzel is a member of several professional organizations, including the Society of Naval Architects and Marine Engineers and the Marine Technology Society, and a director of the Year of the Ocean Foundation. He received B.S. and M.S. degrees in aeronautical engineering from the University of Minnesota. Mr. Wenzel was presented with an honorary doctorate by California Lutheran University for his contributions to ocean engineering.

Lily Y. Young is a professor of environmental microbiology at Rutgers University. Her research interests include microbial physiology and biochemistry, microbial toxicology, and environmental biotechnology. Dr. Young has served on several National Science Foundation and U.S. Environmental Protection Agency advisory and oversight committees on environmental engineering and biodegradation. She has been an invited speaker at many symposia and has published extensively in the area of the microbial degradation of contaminants. She is a fellow of the American Academy of Microbiology and a fellow of the American Association for the Advancement of Science. She has served on the editorial boards of the journals *Applied and Environmental Microbiology* and *Microbial Ecology*. Dr. Young has B.S. and M.S. degrees in bacteriology from Cornell University and a Ph.D. degree in environmental microbiology from Harvard University.

APPENDIX
B

Regulatory Framework for the Management and Remediation of Contaminated Marine Sediments[1]

Kenneth S. Kamlet and Peter Shelley

Numerous federal laws and regulations apply to aspects of the handling and placement of sediments and the means by which they become contaminated. However, no single legal authority is geared specifically to the management of contaminated sediments. Instead, a diverse mix of differing legal requirements comes into play depending on the nature and location of, and the reason for, the dredging and ultimate placement.

ORGANIZATION AND SCOPE

A few words should be said about the coverage and organization of this appendix. It focuses on *contaminated* sediments. However, because contaminated sediments are one category of sediments and many regulatory authorities address sediments in general, this appendix includes a discussion of relevant provisions that primarily concern all types of sediments. For example, provisions of the biennial Water Resource Development Acts (WRDAs) that relate to beneficial uses of dredged material are discussed because beneficial uses are among the management options available for contaminated sediments—despite their more common application in connection with clean sediments. Similarly, this appendix describes the navigation dredging cost-sharing provisions of these acts, even though the provisions do not typically differentiate between contaminated and uncontaminated dredged materials.

---

[1] This appendix has been edited for grammar and style; accuracy and organization are the sole responsibilities of the authors.

Because the committee's report encompasses the broad-based management of contaminated sediment problems, this appendix also includes some evaluation of potentially viable options—using existing, or slightly modified, regulatory tools—for improving control of the sources of sediment contamination. For example, the discussion of applicable Clean Water Act (CWA) authorities is not limited solely to Section 404, which regulates the discharge of dredged and fill material; it also briefly catalogues the CWA provisions that control point source discharges (Section 402), toxics and spills (Sections 307 and 311), and State Water Quality Certification authority (Section 401). Also mentioned are other provisions that are used and, in some cases, could be used more effectively to reduce upstream activities that ultimately impact downstream sediment quality.

On the one hand, the approach taken here could be viewed as an extension of the "systems" or "systems engineering" approach discussed in the report. If the objective is to solve the problem of contaminated sediments, then it is necessary to look at all parts of the regulatory "system" to identify the most workable plan for solving the problem.

On the other hand, the appendix is not intended to be an all-inclusive compilation of environmental laws that may affect a proposal to excavate or dispose of contaminated sediments. Thus, there is only a passing reference to the National Environmental Policy Act of 1969 (NEPA), which applies to all major federal actions that may have a significant environmental impact. And there is no reference at all to the Endangered Species Act (ESA), which can play an important role when sediment handling might disturb imperiled species or their habitat. Although it can be argued that the line drawn is artificial, the authors consider general environmental statutes of this type to be one step removed from sediments and sediment management and therefore not relevant in a survey of authorities governing contaminated sediments.

Admittedly, this appendix is not entirely consistent in this respect. It does include a discussion of the "consistency" provisions of the Coastal Zone Management Act (CZMA),[2] even though the CZMA resembles NEPA and the ESA in that it is a general environmental statute, rather than one geared specifically to sediments. In this case, the authors "erred" on the side of inclusiveness because the CZMA's consistency provisions perform a role very similar to the role of water quality certifications under Section 401 of the CWA. These authorities form the backbone of the legal tools available to coastal states to influence federal regulatory actions in the marine environment. Having decided to address the CWA authorities, the authors decided to include CZMA provisions for completeness.

---

[2] 16 *United States Code*, Section 1451 *et seq.* for the statute and 16 *United States Code*, Section 1456 for the specific provision. References to the *Code* will be abbreviated using the format: 16 USC §1456.

Objectives

Appendix B was organized to accomplish at least five objectives. It supplements and supports the report's references to regulatory and policy issues. It provides the interested reader with a general overview of relevant laws and regulations. It illustrates the different considerations that drive the divergent statutory programs. It provides a sense of the complexity of the regulatory framework and where that framework contains gaps, overlaps, and uncertainties. And it provides some selective indications of how the existing regulatory framework might be changed (without affirmatively recommending specific changes) to enhance its effectiveness or reduce its complexity.

Factors that Drive the Statutory Programs

The regulatory framework discussed in this appendix evolved over many decades. The complexity of this framework is attributable in part to the differing objectives and legal thrusts of the patchwork of statutes and regulations that make it up. Factors that influence which regulatory requirement applies in a given case include the following:

- the navigability of the waterway from which the sediments are excavated—that is, the area in or adjacent to a navigation channel (see, for example, the Rivers and Harbors Act of 1899 [RHA])
- the proposed destination of the sediments—that is, land, ocean, or inland waters (Resource Conservation and Recovery Act [RCRA], CWA, or no law[3] versus Marine Protection, Research and Sanctuaries Act [MPRSA])
- the driving force for sediment management—that is, navigation enhancement (RHA, CWA, and/or MPRSA), environmental remediation/restoration (CWA, Comprehensive Environmental Response, Cleanup, and Liability Act of 1980 [CERCLA],[4] WRDA), water quality improvement (CWA), waste disposal (RCRA, CERCLA), or beneficial use (WRDA)
- the management strategy used—that is, no-action or natural restoration (no law, CERCLA, or CWA[5]), in situ containment or treatment (RHA, CWA, MPRSA, CERCLA, or RCRA), or ex situ containment or treatment (CWA, MPRSA, CERCLA, or RCRA)

---

[3] RCRA may apply to land placement of sediments, especially if there is no "return flow" to waters of the United States and toxic characteristics leaching procedure (TCLP) criteria are exceeded. CWA Section 404 applies if there is a "return flow" to CWA waters. No federal law may apply if the material is contained entirely and TCLP limits are not exceeded.

[4] Superfund is known more formally as the Comprehensive Environmental Response, Cleanup and Liability Act of 1980 (CERCLA), as amended by the Superfund Amendments and Reauthorization Act of 1986 (SARA), 42 USC §§9601–9675.

[5] Restoration could be required in the context of a spill under CWA Section 311.

This patchwork of legislation evolved over several decades as a by-product of the efforts of numerous congressional committees and subcommittees with diverse jurisdictions. In each case, the scope and approach of each statute relate more closely to the jurisdictional authority of the sponsoring congressional committee than to any systematic effort to comprehensively—or even coherently—regulate contaminated sediments.

The appendix is organized into seven sections: (1) the navigation connection (navigation dredging and sediment placement and the role of the RHA and CWA Section 404); (2) site cleanup (remediation and damage restoration provisions of CERCLA); (3) CWA provisions (regulatory provisions other than Section 404); (4) biennial WRDAs (miscellaneous authorizing, regulatory, beneficial use, and funding provisions); (5) state regulatory authorities (CWA Section 401 and CZMA consistency provisions); (6) gaps, overlaps, and uncertainties (including scenarios illustrating how difficult it can be even to determine which authorities apply); and (7) potential regulatory reforms (opportunities for improvement).

THE NAVIGATION CONNECTION

The excavation of sediments requires a U.S. Department of the Army permit under Section 10 of the RHA,[6] when carried out in "navigable" waters. This same statute applies to in-place or ex situ capping, treatment, or subaqueous containment of sediments if the activity has the effect of altering the navigable waterway's "course, location, condition, or capacity."[7] For example, a Section 10 permit is required if sediments are placed in a near-shore or offshore confined aqueous site to create an artificial island or extend waterfront real estate.

Section 10 of the RHA is not an environmental provision; its original purpose was simply to protect the navigable capacity of waterways. However, when an activity for which a permit is required may "significantly affect the quality of the human environment," an environmental impact statement (EIS) may be required under NEPA,[8] which requires the complete assessment and full disclosure of the environmental impacts of, and alternatives to, proposed major federal actions. A full EIS is not required in every instance. The process begins with an environmental assessment (EA). If the EA indicates that there is no significant environmental impact, then the lead agency can make a finding of no significant impact (FONSI), which obviates the need for an EIS. Preparation of a draft and final EIS,

---

[6] Technically, the statute is the River and Harbors Appropriation Act of 1899. Section 10 states: "It shall not be lawful to excavate or fill, or in any manner to alter or modify the course, location, condition, or capacity of . . . any navigable water of the United States, unless the work has been recommended by the [U.S. Army Corps of Engineers]" (33 USC §403).

[7] Id.

[8] 42 USC §§4321–4370.

and associated public and interagency review and comment, can be quite time-consuming—averaging 18 months and often taking several years.

When dredged sediments are "disposed" of in ocean, inland, or near-coastal waters, a U.S. Department of the Army permit is required. For the dumping of dredged material in the ocean (including the territorial sea, which extends three miles out from the mean low water mark), the applicable statutory provision is Section 103 of the MPRSA, popularly known as the Ocean Dumping Act.[9] If the discharge site is in waters of the United States, excluding the territorial sea, then Section 404 of the CWA would apply.[10,11]

Under authority of Section 404, the U.S. Environmental Protection Agency (EPA) develops guidelines in conjunction with the U.S. Army Corps of Engineers (USACE) for specification of dredged or fill material disposal sites. The contaminant status of the material is determined using a manual commonly called the "Gold Book." The Gold Book procedures are used to determine whether the sediment is suitable for unrestricted open-water disposal or whether restriction might be required. The Gold Book is currently being updated.

Section 404 does not prohibit the open-water disposal of highly contaminated sediments as long as management actions, such as capping or treatment, are used to bring the sediment disposal activity into compliance with the guidelines. The use of sediments to create, restore, or enhance wetlands, as well as other beneficial uses that may impact waters of the United States, also are regulated under Section 404 and evaluated using the 404 guidelines.

One potentially troublesome area involves the land placement of contaminated sediments where there is no runoff back into waters of the United States. In these cases, sediment excavation, if in navigable water, would be regulated by Section 10, and the dredged material might be subject to consideration as a hazardous waste under RCRA, if it displayed a hazardous waste "characteristic" (e.g., by TCLP testing). A pending *Federal Register* rule change[12] will address situations in which sediment is proposed for land placement. The proposed rule change does not address whether sediment placement on land is subject to solid-waste regulation by the states. However, the USACE has asserted since at least 1988 that dredged material is not subject to regulation under RCRA, either as a hazardous or a solid waste.[13]

---

[9] 33 USC §§1401–1445.

[10] 33 USC §§1251–1387.

[11] Although it might appear that both the CWA and the MPRSA apply to dredged material discharged to the territorial sea, Congress specified that the MPRSA was to preempt other authorities in the event of an overlap of jurisdiction. It should be noted that, because the MPRSA does not address the placement of "fill material," fill discharges to the territorial sea would be regulated under Section 404 of the CWA.

[12] *Federal Register*, vol. 61, no. 83, April 29, 1996, p. 18849.

[13] *Federal Register*, vol. 53, no. 80, April 26, 1988, pp. 14903, 14910, 14913.

Under authority of MPRSA Section 102, the EPA develops discharge criteria in conjunction with the USACE for the dumping of dredged material in ocean waters. The contaminant status of the material is determined using an ocean dumping manual commonly called the "Green Book." The Green Book procedures determine whether the sediment is suitable for ocean dumping. The latest version of the Green Book was published in February 1991. Green Book procedures are used to determine whether dredged material is acceptable or unacceptable for unrestricted ocean dumping. Before a decision is made regarding dumping in ocean waters, consideration is given to any management actions that may be necessary.

Tiered testing procedures are used, under both the MPRSA and CWA, to evaluate the suitability of dredged sediments for open water placement. These procedures consider the proximity of known pollution sources to the area to be dredged, the physical and chemical properties of the sediments, and, as appropriate, the results of biological tests. For example, under the ocean dumping criteria[14] and associated interagency guidance, a combination of sediment bioassays and bioaccumulation tests is used to assess both the acute toxicity of sediments to resident biota and the potential for the bioaccumulation of sediment contaminants. Based on such tests, dredged material can be classified as either suitable for unrestricted open-water placement or unacceptable for unconfined open-water placement. If the results of laboratory tests indicate a potential for unacceptable adverse effects, then management actions (or management of the placement) need to be considered. Laboratory tests are only indications of potentially unacceptable adverse effects. In making a decision regarding acceptability, the decision maker must consider the effects of the discharge pursuant to 40 CFR §227.13(c)(2)(I) and §227.13(c)(3). Laboratory tests are not pass-fail criteria for purposes of the MPRSA.

There continues to be some debate over legal issues[15] concerning whether there are any circumstances under which dredged material that "fails" the bioassay and bioaccumulation tests can be approved for ocean dumping—even subject to tight management restrictions or under conditions (e.g., placed within geotextile

---

[14] 40 CFR §§223–228.

[15] These issues arise because of an exception in the ocean dumping criteria allowing an applicant to demonstrate that constituents, although present as other than trace contaminants (and, therefore, normally banned from ocean dumping), are: (1) "present . . . only as compounds or forms . . . non-toxic to marine life and non-bioaccumulative in the marine environment upon disposal and thereafter", or (2) "present . . . only as chemical compounds or forms which, at the time of dumping and thereafter, will be rapidly rendered non-toxic to marine life and non-bioaccumulative in the marine environment by chemical or biological degradation in the sea. . . ." (40 CFR §227.6[f]). A federal district court, in *Clean Ocean Action v. York*, 861 F.Supp. 1203 (D.N.J. 1994), reversed on other grounds, 57 F.3d 328 (3rd Cir. 1995), held that neither part of this exclusion would allow post-disposal "capping" of dioxin-contaminated sediments to overcome bioassay results showing toxicity in excess of "trace contaminant" levels. On the other hand, the court was (unaccountably) willing to consider the effects of

bags or covered with a thick cap of uncontaminated sand or clay) designed to ensure the isolation and containment of associated contaminants.

These authorities primarily apply to the placement in open water of large quantities of dredged material. They have no applicability to the in-place treatment or containment of contaminated sediments—except to the extent that other sediments must be "discharged" to "cap" or otherwise contain the contaminants of concern. The authorities are also inapplicable to treatment or containment on land—except to the extent there may be incidental filling of wetlands or other waters.

One of the problems associated with the regulation of dredged material under Section 404 of the CWA is that the emphasis of this program has evolved from specifying open-water disposal sites for dredged material to protecting ecologically valuable wetlands (and other "special aquatic sites"). Thus, many of the procedures (e.g., the required "sequencing" of avoidance, minimization, and mitigation measures, and the need to do detailed "alternatives analysis") mandated under the EPA's 404 guidelines really have little, if any, applicability to the open-water disposal of sediments in connection with the navigation dredging of rivers and harbors. This situation has led to suggestions by, for example, the American Association of Port Authorities, that Section 404 be revised to focus on wetlands (and other "special aquatic sites") and the placement of fill material, with the establishment of a new and separate section to deal with the open-water disposal of dredged material.

SITE CLEANUP

Several hundred million cubic yards of sediments are dredged annually from navigable rivers and harbors. Only a small fraction of this volume can be considered "contaminated" in terms of restrictions on the ability to place the material in open water.[16]

By contrast, many sites with no link to navigation require cleanup for environmental reasons. A growing number of these sites involve significant contaminated sediment problems.

---

capping in evaluating the results of 40 CFR §227.6(c) toxicity testing. The court also held that, even when dredged material is too contaminated to be dumped without capping, the mere fact that there is a loss of 2 to 5 percent of the contaminated sediment in the water column during dumping (i.e., enroute to the bottom) is not a per se violation of the MPRSA (because if the escaping material were unlawful to dump because it can never be capped, no sediment requiring capping could ever be dumped). The EPA plans to revise the ocean dumping criteria to authorize expressly, under specified circumstances, capping of otherwise prohibited material (J. Lishman, EPA, personal communication to K. Kamlet, 1995).

[16] The percentage of the total dredged material that fails the biological tests has increased somewhat in recent years under revised test procedures, which increased the duration of test exposures and mandated the use of more sensitive test organisms.

Superfund

There are approximately 1,300 contaminated sites around the country that are listed or proposed for inclusion on the Superfund National Priorities List (NPL).[17] NPL sites are highly contaminated areas, often associated with prior hazardous-waste disposal activities, that are targeted for priority cleanup through long-term remedial action. Another 10,000 potential Superfund sites are included in an EPA database (the CERCLIS inventory). These sites are assessed systematically by the EPA to determine which ones could be added to the NPL.[18] Many other sites are subject to cleanup under state-level Superfund laws.

According to EPA data,[19] 77 (10.8 percent) of the 712 Superfund NPL sites for which records of decision (RODs) were signed in fiscal years 1982–1991 involved contaminated sediment (both marine and freshwater) as a significantly contaminated "matrix." A much higher percentage of sites (78.5 percent) involved groundwater contamination, based on Superfund's historically greater emphasis on human health than on ecological impacts. Contaminated sediments are likely to be a growing factor at Superfund sites that reach the ROD stage in the future because of the increasing emphasis in EPA regulations on natural resource and food-chain impacts (see footnote 12). An indication of this trend is the large number of NPL sites without RODs—523 representing nearly all of the more than 537 NPL sites not yet at the ROD stage (at this writing)—involving contaminated soil, sediment, or sludge.[20]

In selecting a remedy under Superfund, nine criteria must be addressed:[21]

- overall protection of human health and the environment
- compliance with applicable or relevant and appropriate regulatory requirements (ARARs)

---

[17] Sites are placed on the NPL when they score 28.5 or higher on the EPA's "hazard ranking system." The hazard ranking system is a model that serves as the EPA's primary tool for placing sites on the NPL. The hazard ranking system was amended by Congress in 1986 (SARA, Section 105) to require, for the first time, consideration as a distinct exposure pathway of "the damage to natural resources which may affect the human food chain." Revisions to the hazard ranking system now require consideration of both the water column and contaminated sediments and provide, for the first time, for the placement of a site on the NPL based on environmental contamination as well as human health impacts.

[18] Sites are also removed from the inventory. In 1995, approximately 28,000 sites (classified as "no further remedial action planned") were dropped from the CERCLIS inventory based on preliminary assessments that determined that further remedial action was not required.

[19] U.S. Environmental Protection Agency. 1993. *Cleaning Up the Nation's Waste Sites: Markets and Technology Trends*, EPA 542-R-92-012 (Solid Waste and Emergency Response, OS-110W). Page 16 (Exhibit 2-8), Washington D.C.: EPA.

[20] Environmental Protection Agency. 1993. *Cleaning Up the Nation's Waste Sites: Markets and Technology Trends*, EPA 542-R-92-012 (Solid Waste and Emergency Response, OS-110W). Page 37 (Exhibit 3-9), Washington D.C.: EPA.

[21] 40 CFR §300.430(d).

- long-term effectiveness and permanence
- reduction of toxicity, mobility, or volume through treatment
- short-term effectiveness
- implementability
- cost
- state agency acceptance
- community acceptance

These criteria can be reduced to the following three overall screening factors:

- environmental acceptability (i.e., overall protection of human health and the environment, compliance with ARARs, state agency acceptance, and community acceptance)
- technological feasibility (i.e., long-term effectiveness and permanence; reduction of toxicity, mobility, or volume through treatment; short-term effectiveness; implementability)
- economic viability (i.e., cost)

Where contaminated sediments are addressed under Superfund, the aforementioned nine criteria must be applied in evaluating management alternatives, including compliance with ARARs. ARARs may include "applicable" regulatory requirements, such as "sediment quality criteria" under development by the EPA. They may or may not also include—as urged by representatives of National Oceanic and Atmospheric Administration (NOAA)—"Long and Morgan" sediment values,[22] which are measurements that reflect the contaminant concentrations associated with toxicity to aquatic biota (or maximum nontoxic concentrations) in coastal areas. ARARs also include "relevant and appropriate" regulatory standards, even if not directly applicable in the particular situation. The USACE's dredged material tiered testing procedures, including sediment bioassays and bioaccumulation tests, perhaps could be considered ARARs from this standpoint.

Corrective action levels for soils under RCRA[23] possibly also could be viewed as cleanup standards in appropriate cases. It must be kept in mind, however, that contaminants in dry land soils and contaminants in underwater aquatic sediments have significantly different physical and chemical properties. Thus, regulatory standards established for soils do not have any applicability to sediments.

In addition to Superfund's nine criteria for evaluating remedial alternatives, there is a general statutory preference for treatments that "permanently and significantly reduce the . . . toxicity or mobility" of contaminants (Section 121(b)).

---

[22] E.R. Long and L.G. Morgan. (1991). *The Potential for Biological Effects of Sediment Sorbed Contaminants Tested in the National Status and Trends Program*, NOAA technical memorandum NOS OMA 52. Seattle, Washington: NOAA.

[23] 42 USC §6901 *et seq.*

The in-place or ex situ capping of contaminated sediments generally would not be considered to have the requisite quality of "permanence"—in the same sense as the destruction or detoxification of sediment contaminants would. However, as discussed in the report (e.g., Chapter 5), properly engineered capping may be an optimum management technique for contaminated sediments under some circumstances. Although there may be ways to augment passive capping (e.g., by amending with nutrients or microorganisms to promote biodegradation or by adding activated carbon to physically absorb contaminants) to more closely resemble typical "treatment" methods and thereby gain the status of a "preferred remedy" under Section 121(b), it may be appropriate to consider changing Section 121(b) to clarify the circumstances under which (e.g., low- to intermediate-level contamination spread over large areas of aquatic bottoms) engineered capping would be considered a "preferred" Superfund remedy.

Natural Resource Damage Claims

The Superfund law is known primarily for its provisions relating to the cleanup of NPL sites. However, in addition to these remedial response authorities, the Superfund law contains significant provisions for forcing the restoration of "injured" natural resources. The term "natural resources" is defined expansively to encompass not only fish, animals, birds, and other biota, but also air, water, and groundwater resources.[24] Both the National Contingency Plan (EPA regulations implementing Superfund) and the U.S. Department of the Interior (DOI) natural resource damage assessment regulations define "water resources" to include associated sediments.

Superfund authorizes designated federal and state "trustee" agencies (and Indian tribes) to initiate natural resource damage claims against persons responsible for injury to, destruction of, or loss of natural resources "resulting from" the release or threatened release (causing response costs) of a hazardous substance.[25] Recovered funds must be used "only to restore, replace, or acquire the equivalent of" the injured natural resources, but the measure of damages is not limited by the sums that can be used to restore or replace such resources.[26] Restoration costs can be enormous, particularly in the coastal marine environment, where contaminated sediments can affect hundreds or thousands of square miles. Thus, Superfund remedial actions (which are geared to mitigating an imminent hazard to health or the environment) have an average price tag of $20 million to $25 million per site, a natural resource damage claim (which is geared to restoring the injured resource to its prerelease, undamaged condition) can be orders of magnitude more expensive.

Given that the DOI issued final revised damage assessment regulations in

---

[24] CERCLA §101(16); 42 USC §9601(16).
[25] CERCLA §107(a)(4)(C); 42 USC §9607(a)(4)(C).
[26] 42 USC §9607(f)(1).

APPENDIX B

March 1994,[27] which, from one perspective, started the statute of limitations clock running again, it can be anticipated that a flurry of natural resource damage claims will be brought in the next few years prior to the expiration of the statute.[28] A statute of limitations defines the period of time during which a claimed legal violation or cause of action must be brought. After the statute "expires," such action is barred. In the case of natural resource damage claims under Superfund, the statute of limitations is set at three years from the date of discovery of the resource injury (or the responsible hazardous substance release) or from the date of final promulgation of damage assessment rules, whichever comes later.[29]

Among federal trustee agencies, NOAA is most likely to be concerned with resource damages affecting marine sediments. New York, Texas, Washington, and California have been the coastal states most active to date in pursuing natural resource damage claims.

The primary relevance of Superfund natural resource damage authorities to marine sediment remediation is that they may serve to constrain cleanup options. Specifically, if a contaminated sediment site is part of a natural resource damage proceeding, then mere cleanup or capping to avoid an imminent hazard to health or the environment will not be sufficient. Instead, under Section 9607(f)(1), the only acceptable "remedy" will be one that restores, rehabilitates, or acquires the equivalent of the injured natural resources—including coastal waters, sediments, and associated biota.

Hot Spot and Chronic Sediment Contamination

Superfund sites can involve the contamination of sediments, other environmental media, or both. They also can involve either relatively localized hot spots of contamination or more extensive areas. Often a Superfund "facility" or site will involve multiple "operable units," each requiring distinct types of remedial action. Within an operable unit, there may be identifiable hot spots of contamination toward which the proposed remedy will usually be primarily directed. Other environmental statutes, notably the CWA and various biennial WRDAs, also directly (or indirectly) address the issue of contaminated sediment hot spots.[30]

---

[27] Natural Resources Damage Assessments, Final Rule, *Federal Register*, vol. 59, no. 58, March 25, 1994, p. 14262. These regulations were challenged by numerous industry groups in February 1995 (see footnote 23).

[28] If upheld on appeal, a federal judge's ruling in March 1995 will result in the statute of limitations being treated as expired for any resource injuries discovered more than three years ago. *U.S. v. Montrose Chemical*, CV90-3122 (C.D.Cal. 3/22/95). An appeal was filed immediately with the Ninth Circuit U.S. Court of Appeals.

[29] 42 USC §9612(d)(2).

[30] A summary of applicable provisions of the EPA's major environmental statutes, some of which are not addressed in detail in this review, can be found in Environmental Protection Agency. 1991. Contaminated Sediments: Relevant Statutes and EPA Program Activities, EPA 506/6-90-003. (Table 1). Washington D.C.: EPA.

CLEAN WATER ACT PROVISIONS

The CWA contains a number of provisions—in addition to Section 404, which regulates the discharge of dredged or fill material into navigable waters of the United States—that have general or site-specific relevance to contaminated sediments and sediment-associated contamination.

CWA Section 115[31]

CWA Section 115 (in-place toxic pollutants), although seldom funded and even less frequently utilized,[32] directs the EPA "to identify the location of in-place pollutants with emphasis on toxic pollutants in harbors and navigable waterways." This section also authorizes the EPA, acting through the USACE, "to make contracts for the removal and appropriate disposal of such materials from critical port and harbor areas." Thus, Section 115 takes the common sense approach of allowing the removal of hot spots of toxic pollutants to be "piggy-backed" on nearby dredging carried out for navigation reasons, with the EPA reimbursing the USACE for the incremental costs. The eminently plausible logic supporting this approach is that dredging for hot spot removal in conjunction with a navigation project would be far less costly than carrying out a separate, free-standing remediation project. Note that "appropriate disposal" still is required for any contaminated sediments that are so excavated. It is unfortunate that this provision, which is the only CWA provision directed at managing in-place contaminated sediments, is not used more often.

CWA Section 303

CWA Section 303 (water-quality-based discharge limits)[33] requires each state to establish numerical or narrative water quality standards for each pollutant to protect designated uses of regulated waterway segments. However, the CWA's primary mechanism for controlling point sources of water pollution is the use of technology-based effluent limits. States must identify water-quality-limited bodies of water—those that cannot meet the quality-based standards simply by adhering to technology-based limits.

Where technology limits prove insufficient to meet water quality standards in a given waterway segment, more stringent, water-quality-based discharge limits are required to be imposed under Section 303. After a state establishes and the EPA approves a list of quality-limited waterway segments, the state must conduct a study to establish the total maximum daily load (TMDL) for each pollutant that

---

[31] 33 USC §1265.

[32] The EPA indicates that CWA Section 115 was funded once in 1977 and applied to the Duwamish Waterway in Puget Sound.

[33] 33 USC §1313.

the body of water can receive without violating the water quality standard. The TMDLs then are used to establish waste load allocations (WLAs) for each point source of each pollutant, after leaving unallocated a portion of the TMDL as a margin of safety. Nonpoint sources, such as storm water runoff, are assigned load allocations (LAs). As one of the CWA's few watershed-oriented regulatory provisions, and one that addresses both point and nonpoint sources of pollution, this provision has a significant potential to control upstream sources of downstream sediment contamination.

Although the requirements for TMDLs, WLAs, and LAs have been part of the CWA since 1972 (and the EPA's list of pollutants requiring water-quality-based limits was promulgated in 1978), the EPA and the states have been unable to implement this program fully. However, environmentalists have shown a recent willingness to use CWA citizen suit authority to require the EPA to enforce more vigorously state adherence to Section 303 TMDL requirements.[34] In the absence of new legislation explicitly requiring pollutants to be managed on a watershed-wide basis, or expressly holding upstream discharge sources accountable for resultant downstream sediment contamination,[35] water quality discharge

---

[34] Recent court cases may be of interest to ports or others looking for a legal mechanism to force cleanup of sources. *Alaska Center for the Environment* v. *Reilly*, 796 F.Supp. 1374 (W.D. Wash. 1992), aff'd sub nom. *Alaska Center for the Environment* v. *Browner*, 1994 WL 101029 (9th Cir. March 30, 1994) (the EPA held to have a nondiscretionary duty to establish TMDLs where a state failed to do so, in response to a citizen suit by a coalition of environmental groups to force the EPA to adopt TMDLs in Alaska, where the state had failed to do so); *Dioxin/Organochlorine Center* v. *Rasmussen*, 1993 W.L. 484888 (W.D. Wash. August 10, 1993) (upheld against an industry and environmental group challenge to an EPA TMDL and waste load allocation for dioxin set for the Columbia River Basin in Washington, Oregon, and Idaho, despite the lack of previously established technology-based effluent limits for dioxin); *Sierra Club, North Star Chapter* v. *Browner*, 843 F.Supp. 1304 (D.Minn. 1993) (rejected citizen suit after the EPA had acted to reject Minnesota's list of water-quality-limited segments and gave EPA broad latitude to set reasonable timetables for developing TMDLs); and *American Paper Institute* v. *EPA*, 996 F.2d 346 (D.C.Cir. 1993) (upheld an EPA rule requiring discharge permits to contain specific limits, even if the water quality standards on which they are based are solely narrative rather than numerical).

[35] The Japanese approach to distant contributors to contamination problems that interfere with important public works projects is instructive—although, apparently, not widely implemented. Under Japan's Pollution Control Public Works Cost Allocation Law (Law No. 133 of December 25, 1970, as amended by Law No. 43 of 1978), businesses that "engage in . . . industrial activities which cause or will cause pollution" in the area where specified "pollution control public works" take place, must assume financial responsibility for an "amount proportionate to the degree that the industrial activities of all the enterprises . . . constitute the source of pollution for which [the] pollution control public works are undertaken." (Articles 3 and 4). Among the "public works" projects to which this law applies are "[d]redging . . . or any other works as prescribed by a Cabinet Order, which are undertaken in rivers, lakes, harbours, or any other area for public use where sludge and other pollution causing substances are deposited or where water is polluted." (Article 2.2(2)). Port authorities are given the status of a "local government body" for purposes of implementing, enforcing, and receiving reimbursement under this law (Articles 11, 19). The closest existing counterparts under U.S. law are the "joint and several liability" provisions under Superfund.

limits under Section 303 are among the few available mechanisms under existing law by which regulators could more aggressively regulate upstream pollution sources that impact sediment quality.

CWA Section 304(l)

CWA Section 304(l) (toxic hot spots)[36] required the states to identify in 1989 those state waters that could not attain or maintain ambient water quality standards "due to toxic pollutants." For each segment of toxics-limited navigable waters, the states were to identify "the specific point sources" discharging any toxic pollutant and the amount of each such pollutant believed to be contributing to such water quality impairment. Finally, the states were to develop an "individual control strategy" for each waterway segment capable of meeting applicable water quality standards within three years. This provision is not directed at contaminated sediments.[37] However, to the extent it promotes the control of point-source toxic discharges that contribute to sediment contamination, it is a relevant component of a contaminated sediments remediation strategy. Although the statutory deadlines for state action under this provision have passed, the provision directs the EPA to implement the requirements in the absence of an approvable state strategy.

CWA Section 307

CWA Section 307 (toxic pollutants and pretreatment)[38] requires effluent limitations based on the "best available technology economically achievable" for the applicable category or class of point sources to be applied to discharges of specified priority toxic pollutants. This section also requires that pretreatment standards be applied to prevent discharges into publicly owned treatment works from interfering with, passing through, or otherwise being incompatible with such treatment works. Like Section 304(l), this section is of indirect relevance to the problem of contaminated sediments and toxic hot spots.

CWA Section 319

CWA Section 319 (nonpoint-source pollution)[39] requires states to submit to the EPA for approval a report identifying navigable waters within the state that, without additional action to control nonpoint sources, could not reasonably be

---

[36] 33 USC §1314(l).

[37] The EPA notes that, although this provision was not directed at contaminated sediments, a number of sites identified contaminated sediments as the primary source of toxic pollutants to the water column.

[38] 33 USC §1317.

[39] 33 USC §1329.

expected to attain or maintain applicable water quality standards or goals. States then are required to establish and implement a management program for nonpoint sources, emphasizing a watershed approach and using "best management practices and measures." Although nonpoint-source pollution contributes, along with point source discharges, to downstream sediment contamination, it is somewhat less likely than discrete point sources to contribute to hot spots of sediment contamination.[40]

CWA Section 320

CWA Section 320 (National Estuary Program)[41] provides for the nomination and designation of estuaries of national significance that will be subject to supplemental controls on point and nonpoint sources of pollution based on a comprehensive management plan for the estuary. Sixteen estuaries throughout the United States were to be given priority consideration under this program. Most of these estuaries have since been addressed by the program. Section 320 is one of a number of geographically specific provisions under the CWA and WRDAs that may result in a variety of additional contaminated sediment management requirements at specific locations.

CWA Section 311

CWA Section 311 (oil and hazardous substance spills)[42] establishes strict liability for discharges of oil or hazardous substances into or upon the navigable waters of the United States. This section supplements the provisions addressing point and nonpoint sources by focusing on contaminants introduced by inadvertent spills.

CWA Section 402

CWA Section 402 (point source discharges)[43] establishes a permit program for point source discharges of pollutants into waters of the United States. This program has been delegated to the states in most parts of the country. In terms of contaminated sediments, one of the major limitations of this program is that it has

---

[40] A recent EPA background report (EPA. 1996. National Sediment Contaminant Source Inventory: Analysis of Facility Release Data, EPA-823-D-96-001. Office of Science and Technology. Washington D.C.: EPA), which relied on Toxic Release Inventory and Permit Compliance System data (but did not look at nonpoint-source pollution), provides preliminary evidence that active industrial discharges may be an ongoing contributor to sediment contamination and that sediment contamination is not strictly a hot spot issue associated with historical discharges.

[41] 33 USC §1330.

[42] 33 USC §1321.

[43] 33 USC §1342.

tended to address, for any given discharge source, only a few of the toxic pollutants that may be present and may contribute to sediment contamination.[44] This limitation is compounded by the fact that, once a discharge permit is issued, it "shields" the discharger against later abatement efforts or damage claims directed even at pollutants not specifically addressed in the permit. Some EPA regions and states have sought to address this problem by requiring, in certain instances, the use of whole effluent bioassays and biomonitoring as permit conditions, so that the focus is placed on reducing overall toxicity—regardless of the mix of pollutants contributing to the toxicity.

CWA Section 118(c)(3)

CWA Section 118(c)(3) (toxics in Great Lakes sediments)[45] established a five-year study and demonstration program in the Great Lakes relating to the control and removal of toxic pollutants, with an emphasis on bottom sediments. The EPA was directed to publish a variety of information, including "specific numerical limits" to protect health, aquatic life, and wildlife from the bioaccumulation of toxic substances. A final report to Congress was due at the end of 1993. This was the basis for the EPA's Assessment and Remediation of Contaminated Sediments (ARCS) program. The ARCS program has evaluated a number of contaminated sediment technologies—particularly at heavily contaminated sites in the U.S. Great Lakes, known as "areas of concern" under the U.S.-Canada Great Lakes Water Quality Agreement.

Other CWA Provisions

The CWA identifies a number of other specific regional problem areas for targeted remedial or planning efforts related to contaminated sediment issues. These areas include the Chesapeake Bay (establishment of a Chesapeake Bay Program to determine, among other things, the impact of sediment deposition in the bay and the sources, rates, routes, and distribution patterns of such sediment deposition)[46]; the upper Hudson River (project to demonstrate methods for the selective removal of polychlorinated biphenyl (PCBs) contaminating bottom sediments of the Hudson River)[47]; and Long Island Sound (management study to address issues including "contaminated sediments and dredging activities").[48]

---

[44] The EPA indicates that the National Pollutant Discharge Elimination System program, which has responsibility for the issuance of discharge permits under Section 402, has now "agreed to use the EPA standardized bioassays to evaluate sediment contamination."
[45] 33 USC §1268.
[46] CWA Section 117, 33 USC §1267.
[47] CWA Section 116, 33 USC §1266.
[48] CWA Section 119, 33 USC §1269.

BIENNIAL WATER RESOURCE DEVELOPMENT ACTS

Since the nineteenth century, Congress has periodically enacted public works legislation authorizing water resource projects. Originally termed the Rivers and Harbors Acts, which were adopted at irregular intervals, more recent legislation has been enacted as Water Resource Development Acts, and there has been an effort to enact them every two years (there was a gap in 1994). Although WRDA statutes are primarily intended to authorize federal funding for particular navigation and flood control projects and to specify cost-sharing formulas for eligible projects, they have often been used as vehicles for modifying the regulatory framework (or for waiving or varying certain regulatory requirements on a project-specific basis) for water resource projects. The following section summarizes provisions of the various WRDA statues that either directly address sediment issues or deal more broadly with management issues discussed in the report. These issues include cost-sharing for navigation projects and incentives for the beneficial use of dredged material.

WRDA 1986

WRDA 1986[49] reformed the USACE's Civil Works program by establishing a comprehensive cost-sharing scheme for distributing the construction costs for water resource development projects between the U.S. government and nonfederal interests.[50] The percentage of the nonfederal contribution for navigation projects depends on the depth of the project. For all navigation projects, however, the act requires that nonfederal interests (i.e., local sponsors) provide all necessary lands, easements, rights-of-way, and dredged material placement areas, as well as perform necessary operation and maintenance (this is sometimes referred to as "the local cooperation requirement").[51] The value of these contributions is credited toward the nonfederal interest's share of the project costs. A 1993 Department of the Army legal opinion[52] held that the local sponsor is also responsible (in most cases) for "any diking costs necessary to prepare a site to function as a disposal area."[53] (WRDA 1986 also imposed a harbor maintenance tax and an inland

---

[49] P.L. 99-662, 100 Stat. 4082.

[50] Sections 101–109, 100 Stat. 4082–4089, 33 USC §2211 (1988).

[51] Sections 101(a)(3), 101(e)(1), 100 Stat. 4083. See also, 33 USC §2211, 42 USC §1962d–5b.

[52] July 23, 1993, memorandum for the director of civil works from G. Edward Dickey, acting assistant secretary of the army (civil works), on "confined disposal facilities (CDF)," transmitting a June 29, 1993, memorandum from Earl H. Stockdale, deputy general counsel (civil works and environment).

[53] 33 USC §419a (1988), however, directs the secretary of the army to "utilize and encourage the utilization of such management practices as he determines appropriate to extend the useful life of dredged material disposal areas." Also, Section 216 of WRDA 1992 authorizes the secretary to conduct a study "on the need for changes in Federal law and policy with respect to dredged material disposal areas for the construction and maintenance of harbors and inland harbors by the Secretary"— including "the need for any changes in Federal and non-Federal cost sharing for such areas and harbor projects, including sources of funding." WRDA 1992 §216(a), 106 Stat. 4832–4833 (1992).

waterways tax and authorized the creation of a Harbor Maintenance Trust Fund and an Inland Waterways Trust Fund;[54] money from the former trust fund was to be used to fund up to 40 percent of eligible operations and maintenance costs assigned to the commercial navigation of all harbors and inland harbors in the United States.[55] Nonfederal interests to generate funds to cover their cost shares, are authorized to levy port or harbor dues on vessels and cargo utilizing a harbor.[56])

One of the ironic (and probably unintended) consequences of making the local sponsor responsible for dredged material placement facilities is that, all else being equal, a strong economic incentive is created to use open-water sites (which are "free") in preference to land and near-shore sites, which must be paid for by the project proponent. It also creates a dichotomy between ports and harbors in different parts of the country. For example, Section 123 of the Rivers and Harbors Act of 1970 authorizes the secretary of the army to construct, operate, and maintain contained disposal facilities (CDFs) in the Great Lakes and their connecting channels, with local interests generally bearing none of the costs and, under certain circumstances, bearing only 25 percent of the construction costs.[57]

Because the presence of contaminated sediments in an area that lacks adequate, environmentally appropriate placement capacity for the material can create major impediments to proceeding with commercially essential navigation dredging, the issue of who must construct and operate a CDF can be of critical significance. There is a proposal in WRDA 1996 to require the USACE to contribute to the cost of land placement facilities, including CDFs for dredged material, on the same cost-sharing basis as specified in WRDA 1986 for new-work dredging.

Other WRDA 1986 provisions of interest include:

- Section 201(a), dealing with deep-draft harbor development projects authorizes the creation of 800 acres of land with dredged material from deepening of the entry channels to the harbors of Los Angeles and Long Beach, California;[58] directs that the disposal of beach quality sand from the deepening of New York Harbor and adjacent channels in New York and New Jersey shall take place at specified oceanfront beaches "at full federal expense" and prohibits the placement of dredged material from these projects at Bowery Bay, Flushing Bay, Powell's Cove, Little Bay, or Little Neck Bay[59]; requires the USACE, in connection with the Oakland Outer Harbor,

---

[54] Sections 1401–1405, 100 Stat. 4266–4272.
[55] Section 210, 100 Stat. 4106. See also, 33 USC §2238, 26 USC §9505.
[56] 100 Stat. 4102–4106, 33 USC §2236.
[57] 33 USC §1293a (1988).
[58] 100 Stat. 4091.
[59] 100 Stat. 4091.

APPENDIX B   199

California, navigation project, to "study alternative dredged material disposal plans, including . . . plans which include marsh formation" and to monitor the effects of dredged material placement measures, including "such measures as will result in fish and wildlife habitat enhancement"[60]; directs the USACE, in connection with the Duluth-Superior, Minnesota and Wisconsin, navigation project, to study "whether it would be more cost-effective and environmentally sound to control future sedimentation than to conduct periodic maintenance dredging of such project"[61]; and directs, in connection with the Gulfport Harbor, Mississippi, navigation project, that, "for reasons of environmental quality, dredged material from such project shall be disposed of in open water in the Gulf of Mexico in accordance with all provisions of Federal law" and that, for purposes of economic evaluation, "the benefits from such open water disposal shall be deemed to be at least equal to the costs of such disposal."[62]

- Section 211 directed the EPA to designate within three years one or more alternative dredged material ocean dump sites, "not less than 20 miles from the shoreline," for disposal of dredged material currently placed at the Mud Dump site. Following the designation of new site(s), only "acceptable dredged material" (defined as "rock, beach quality sand, materials excluded from testing under the ocean dumping regulations. . . , and any other dredged material [including that from new work] determined by the Secretary, in consultation with the Administrator, to be substantially free of pollutants") could continue to be placed at the Mud Dump.[63]

- Section 704 directed the secretary of the army to investigate and study the feasibility of using the capabilities of the USACE "to conserve fish and

---

[60] 100 Stat. 4092.

[61] 100 Stat. 4094.

[62] 100 Stat. 4094–4095.

[63] 100 Stat. 4106–4107. This provision was opposed by both the dredging community and environmentalists and has yet to be implemented—although studies to identify and designate an alternative ocean dump site continue. WRDA 1988, Section 32, required the EPA administrator (by mid-March of 1989) to submit a site designation plan, including a specific schedule with milestones. 102 Stat. 4030–4031. WRDA 1990, Section 412, finally repealed the 1986 provision and instead requires the EPA and the USACE (within 180 days) to submit "a plan for the long-term management of dredged material from the New York/New Jersey Harbor region." 104 Stat. 4650. This plan was to include a discussion of potential alternative placement sites, including the feasibility of altering the boundaries of the Mud Dump; measures to reduce the quantities of dredged material proposed for ocean dumping; and measures to reduce the amount of contaminants in materials proposed to be dredged from the harbor through source controls and decontamination technology. The USACE, in consultation with the EPA, was directed to implement a demonstration project for placement of up to 10 percent of the material dredged from the harbor region "in an environmentally sound manner other than by ocean disposal"—including, among others, "capping of borrow pits, construction of a containment island, application for landfill cover, habitat restoration, and use of decontamination technology."

wildlife (including their habitats)."[64] This could result in an expansion of USACE authority to allow the beneficial use of dredged material for the conservation of fish and wildlife—for example, by using dredged material to create wetlands or other wildlife habitat.

- Section 709 directed the EPA to "study and monitor the extent and adverse environmental effects of dioxin contamination in the Passaic River-Newark Bay navigation system," with a report back to Congress including recommendations "concerning methods of reducing the effects of such contamination."[65]
- Section 730 directed the secretary of the army to study "current practices on the sharing of costs related to the benefits of increased land values resulting from water resources projects [carried out by the USACE], together with potential methods by which any increase in land values should be shared between the Federal Government and the non-Federal interests."[66]
- Section 904 requires the USACE to display in the benefits and costs of water resource projects "the quality of the total environment, the well-being of other people of the United States, the prevention of loss of life, and the preservation of cultural and historical values. . . ."[67]
- Section 906 authorizes the USACE to carry out "mitigation of fish and wildlife losses, including the acquisition of lands or interests in lands to mitigate [such] losses, as a result of [a water resource] project," with the "first costs" of such enhancements to be a federal cost when the benefits are determined to be national (otherwise, nonfederal interests must pay 25 percent of these first costs).[68]
- Section 907 specifies that, in evaluating the benefits and costs of a water resources project, the benefits attributable to measures included for the purpose of environmental quality enhancement "shall be deemed to be at least equal to the costs of such measures."[69]
- Section 908 establishes an Environmental Protection and Mitigation Fund to pay the federal share of mitigation costs.[70]
- Section 1135 authorizes the secretary to review the operation of completed water resource projects to determine the need for modifications "for the purpose of improving the quality of the environment in the public interest."[71]

---

[64] 100 Stat. 4157, 33 USC §2263.
[65] 100 Stat. 4159.
[66] 100 Stat. 4165.
[67] 100 Stat. 4185, 33 USC §2281.
[68] 100 Stat. 4186–4187, 33 USC §2283.
[69] 100 Stat. 4188, 33 USC §2284.
[70] 100 Stat. 4188, 33 USC §2285.
[71] 100 Stat. 4251–4252.

- Section 1162 directed the USACE "to remove polluted bottom sediments" from the Miami River and Seybold Canal, Florida, with local interests furnishing all lands, easements, rights-of-way, relocations, and alternations necessary for initial dredging and subsequent maintenance.[72]

WRDA 1988

WRDA 1988 contains only a few provisions of interest:

- Section 4(n)[73] authorizes the USACE to place dredged material from the Gulfport Harbor, Mississippi, navigation project "in accordance with all provisions of Federal law" at various open-water locations, including "thin layer disposal" in the Mississippi Sound of new-construction dredged material as part of a demonstration project. It also directs the USACE to carry out a comprehensive demonstration program "for the purpose of evaluating the costs and benefits of thin layer disposal. . . and for determining whether or not there are unacceptable adverse effects from such disposal." "Thin layer disposal" is defined as "the deliberate placement of a 6- to 12-inch layer of dredged material in a specific bottom area."
- Section 8[74] directs the secretary of the army, "whenever feasible, [to] seek to promote long- and short-term cost savings, increased efficiency, reliability, and safety, and improved environmental results through the use of innovative technology in all phases of water resources development projects and programs under the Secretary's jurisdiction." Such measures include encouraging "greater participation by non-Federal project sponsors in the development and implementation of projects." "Innovative technology" is defined as "designs, materials, or methods which the Secretary determines are previously undemonstrated or are too new to be considered standard practice."
- Section 24[75] amends Section 123 of the Rivers and Harbors Act of 1970 to authorize the USACE "to continue to deposit dredged materials into a contained soil disposal facility constructed under this section [i.e., in the Great Lakes] until the Secretary determines that such facility is no longer needed for such purpose or that such facility is completely full." The USACE is directed to "conduct a study of the materials disposed of in [such CDFs] . . . for the purpose of determining whether or not toxic pollutants are present in such facilities [and their concentrations]." "Toxic

---

[72] 100 Stat. 4257–4258.
[73] 102 Stat. 4017–4019.
[74] 102 Stat. 4023–4024, 33 USC §2314.
[75] 102 Stat. 4027–4028.

pollutant" is defined as in CWA Section 301(b)(2) and "such other pollutants as the Secretary, in consultation with the Administrator . . . determines are appropriate based on their effects on human health and the environment."

WRDA 1990

- Section 312[76] authorized the USACE, as part of the operation and maintenance of navigation projects, "to remove . . . contaminated sediments . . . outside the boundaries of and adjacent to [a] navigation channel" whenever necessary to meet CWA water quality requirements. The USACE was further authorized to remove contaminated sediments from navigable waters "for the purpose of environmental enhancement and water quality improvement" when requested to do so by a nonfederal sponsor and the sponsor agrees to pay 50 percent of the removal cost. Such removal need not be associated with any navigation project. Disposal costs for all contaminated sediment removal under this section are declared to be a "non-Federal responsibility," and the law in no way affects any party's liability under Superfund (CERCLA).[77] Although only $10 million annually was appropriated for such purposes, this "environmental dredging" authority represented an important new dimension to the USACE activities.
- Section 306[78] directed the secretary of the army to "include environmental protection as one of the primary missions of the USACE in planning, designing, constructing, operating, and maintaining water resources projects."
- Section 307[79] established, as part of the USACE's water resources development program, "an interim goal of no overall net loss of the Nation's remaining wetlands base, as defined by acreage and function, and a long-term goal to increase the quality and quantity of the Nation's wetlands, as defined by acreage and function." It also directed the secretary to develop, in consultation with other agencies, a "wetlands action plan" to achieve these goals as soon as possible. The secretary also was authorized, in consultation with the administrator, to "establish and implement a demonstration program for the purpose of determining the feasibility of wetlands restoration, enhancement, and creation. . . ." This provision is relevant because of the role of sediments in projects to restore, enhance, and create wetlands.
- Section 401[80] authorized the secretary to provide technical, planning, and

---

[76] 104 Stat. 4639–4640.
[77] P.L. 101-640, Title III, §312 is set forth at Statutory Notes at 33 USC §1252 (1993 Supp.).
[78] 104 Stat. 4635, 33 USC §2316.
[79] 104 Stat. 4635–4637, 33 USC §2317.
[80] 104 Stat. 4644.

engineering assistance to states and local governments in developing and implementing "remedial action plans for areas of concern in the Great Lakes identified under the Great Lakes Water Quality Agreement of 1978." Nonfederal interests must contribute 50 percent of the costs of such assistance.

- Section 411[81] directed the assistant secretary of the army for civil works, the EPA administrator, and the governor of New York to jointly convene "a management conference for the restoration, conservation, and management of Onondaga Lake, New York" to develop recommendations for priority corrective actions and compliance schedules for cleanup and coordinate implementation of the plan. Administrative services are to be provided by a new not-for-profit corporation, and the USACE and EPA are authorized to provide 70 percent grants to the state for the discharge of its responsibilities under this provision. Among the allowable uses of grant funds is "gathering data and retaining expert consultants in support of litigation undertaken by the State of New York to compel cleanup or obtain cleanup and damage costs from parties responsible for the pollution of Onondaga Lake. . . ." Because Onondaga Lake is the subject of active litigation over natural resource damages under CERCLA, the provision specifies that grants made under this section "shall not relieve from liability any person who would otherwise be liable under Federal or State law for damages, response costs, natural resource damages, restitution, equitable relief, or any other relief."

WRDA 1992

WRDA 1992[82] contains numerous provisions dealing with the management of contaminated sediments, the beneficial uses of dredged material, and dredged material management options. This act includes the "National Contaminated Sediment Assessment and Management Act" as Title V of WRDA 1992.[83] Some of the most important provisions of Title V are discussed here.

"Contaminated sediment" is defined as aquatic sediment that "contains chemical substances in excess of appropriate geochemical, toxicological or sediment quality criteria[84] or measures" or "is otherwise considered by [EPA] to pose a threat to human health or the environment." (Section 501(b)(4)).

---

[81] 104 Stat. 4648–4650.
[82] 106 Stat. 4797.
[83] Sections 501–510, 106 Stat. 4864–4871.
[84] The EPA has developed and promulgated five sediment quality criteria (for acenaphthene, dieldrin, endrin, fluoranthene, and phenanthrene) along with the technical basis for establishing sediment quality criteria for non-ionic chemicals using equilibrium partitioning. See, *Federal Register*, vol. 59, no. 11, January 18, 1994, pp. 2652–2656.

A National Contaminated Sediment Task Force was to be established and to include representatives of the EPA, the USACE, NOAA, the U.S. Fish and Wildlife Service, U.S. Geological Survey (USGS), U.S. Department of Agriculture; up to three state representatives; up to three representatives of ports, agriculture, and manufacturing; and up to three representatives of public interest organizations. The task force, which was to submit a report to Congress by November 1, 1994, was to review reports on the extent and seriousness of aquatic sediment contamination in the United States; review programs on contaminated sediment restoration methods, practices, and technologies; review the selection of pollutants for the development of aquatic sediment criteria; provide advice on the development of guidelines for contaminated sediment restoration; recommend practices and measures to prevent contamination of aquatic sediments and control sources of sediment contamination; and "review and assess the means and methods for locating and constructing permanent, cost-effective long-term disposal sites for the disposal of dredged material that is not suitable for ocean dumping. . . ."[85] (Section 502).

The EPA, in consultation with NOAA and the USACE, was directed to "conduct a comprehensive national survey of data regarding aquatic sediment quality in the United States."[86] The survey was to include a compilation of "all existing information on the quantity, chemical and physical composition, and geographic location of pollutants in aquatic sediment, including the probable source of such pollutants and identification of those sediments which are contaminated. . . ." The resulting report to Congress (due by November 1, 1994) was to include "recommendations for actions necessary to prevent contamination of aquatic sediments and to control sources of contamination." (Section 503(a)).[87]

---

[85] Creation of such a task force was one of the recommendations in a 1989 National Research Council report, Contaminated Marine Sediments: Assessment and Remediation. Washington, D.C.: National Academy Press. The EPA indicated that the federal agency participants in the task force were expected to have their first meeting by the end of 1994.

[86] This was also one of the recommendations in a 1989 NRC report, Contaminated Marine Sediments: Assessment and Remediation. Washington, D.C.: National Academy Press.

[87] The EPA's Office of Science and Technology (OST) has been working since 1992 to develop this inventory of contaminated sediment sites. Based on experience gained from pilot inventories in EPA Regions IV, V, and the Gulf of Mexico Program (a combination of Regions IV and VI) during 1992 and 1993, a planning document, "Framework for the Development of the National Sediment Inventory," was produced in November 1992. This document describes the approach to be used for sediment quality data and discusses how the information will be used by each EPA program office. The document was presented and discussed at a two-day interagency workshop held in Washington, D.C., in March 1993. For the next year, OST compiled data from more than 10 national and regional data sets into a centralized database called the National Sediment Inventory (NSI). The NSI is being evaluated to produce the first report to Congress on sediment quality in the United States. Also included in this report to Congress will be data from a two-year study of point-source discharges of sediment contaminants nationwide, including an analysis of areas, chemicals, and industries of concern. A

The EPA, in consultation with NOAA and the USACE, was also directed "to conduct a comprehensive and continuing program to assess aquatic sediment quality"—including an assessment of aquatic sediment quality trends over time and the establishment of a clearinghouse for information on technology, methods, and practices available for the remediation, decontamination, and control of sediment contamination. The initial report was due by November 1, 1994, with updated reports due biennially thereafter (Section 503(b)).

The MPRSA was amended in a number of respects. Changes included (1) changes to Section 103(c) requiring EPA concurrence in all USACE permit actions and precluding issuance of an ocean dumping permit by the USACE if the administrator declines to concur; (2) changes to Section 106(d) eliminating federal preemption of the right of states "to adopt or enforce any requirements" with respect to the dumping of materials "into ocean waters within the jurisdiction of the State" (i.e., usually out to the three-mile limit of the territorial seas); (3) changes to the site designation process under Section 102(c), including development by the EPA in conjunction with the USACE of a "site management plan," updated at least every 10 years, for each designated dredged material placement site (must include a monitoring program, special management conditions, "consideration of the quantity of the material to be disposed of at the site, and the presence, nature, and bioavailability of the contaminants in the material," and the anticipated closure date for the site); (4) a prohibition on final designation of a site after January 1, 1995, unless a management plan has been developed for the site; (5) a prohibition on issuing a permit for dumping at a site after January 1, 1997, unless the site has received final designation; (6) for previously designated sites, management plans must be completed no later than January 1, 1997; and (7) elimination of the preexisting practice under Section 103(b) of easy and open-ended "selection" of dumping sites by the USACE in the absence of EPA-designated sites (USACE selections now can occur only for a maximum of 10 years and only if no feasible placement site has been designated by the EPA; continued use of an alternative site is necessary to maintain navigation and commerce; and continued use of the site does not pose an unacceptable risk to health, aquatic resources, or the environment) (Section 506).

Section 203[88] authorizes the USACE to "accept contributions of cash, funds, materials and services" from entities other than the project sponsor to assist in

---

preliminary evaluation of the sediment chemistry portion of the NSI, which identifies each U.S. watershed area with elevated chemical concentrations that may pose an ecological and/or human health risk, is described in an EPA report, *The National Sediment Inventory: Preliminary Evaluation of Sediment Chemistry Data*, dated May 17, 1994. A second interagency workshop was held in April 1994 to identify a methodology to be used in evaluating all of the NSI data so as to identify both contaminated sediment sites and chemicals of concern. After application of this methodology, the final report to Congress was expected to be completed in the spring of 1995.

[88] 33 USC §2325, 106 Stat. 4826.

carrying out "a water resources project for environmental protection and restoration or . . . recreation. . . ." This recognizes that third parties may benefit from and be willing to contribute to environmental protection projects.

Section 204[89] authorizes the USACE to undertake "projects for the protection, restoration, or creation of aquatic and ecologically related habitats" where the "monetary and nonmonetary" benefits of the project justify its cost, and the project does not result in environmental degradation. Nonfederal interests must enter into a cooperative agreement to provide 25 percent of the incremental construction cost, "including provision of all lands, easements, rights-of-way, and necessary relocations," and 100 percent of "the operation, maintenance, replacement, and rehabilitation costs" associated with the habitat enhancement project. This provision was designed to establish an outlet for the beneficial use of dredged material. The USACE is authorized to spend up to $15 million annually on projects under this section.

Section 207[90] amends WRDA 1976, Section 145, by authorizing the USACE to enter into agreements directly with political subdivisions (at the request of a state) to use dredged material for beach nourishment. It also encourages the USACE to accommodate the state's schedule for paying its share of the cost.

Section 216[91] directed the USACE to "conduct a study on the need for changes in Federal law and policy with respect to dredged material disposal areas for the construction and maintenance of harbors and inland harbors by the [USACE]." Specifically, this study was to "evaluate the need for any changes in Federal and non-Federal cost sharing for such areas and harbor projects, including sources of funding." A report, with recommendations, was to be submitted to Congress by the spring of 1994.

Section 308[92] has potential relevance to the Port of Baltimore (addressed in the Hart and Miller islands case history in Appendix C of this report) because of the treatment under Maryland law of all [dredged] "spoil" from Baltimore Harbor as presumptively contaminated. It called for a study of Baltimore Harbor (with a report due by April 1, 1993) by the USACE "for the purpose of developing analytical procedures and criteria for contaminated dredged material in order to distinguish those materials which should be placed in containment sites from those materials which could be used in beneficial projects (such as beach nourishment, shoreline erosion control, island reclamation, and wetlands creation) or which could be placed in open waters without being chemically altered."[93] This section

---

[89] 33 USC §2326, 106 Stat. 4826–4827.

[90] 33 USC §426(j), 106 Stat. 4829.

[91] 106 Stat. 4832–4833.

[92] 106 Stat. 4841–4842.

[93] This is doubtless in response to a provision of Maryland law that defines, by legislative fiat, all material dredged from Baltimore Harbor as "polluted" and subject to containment at the Hart and Miller islands diked containment facility. This state requirement has resulted in the very inefficient use of limited disposal capacity at the Hart and Miller site. It also has prevented the implementation of desirable beneficial use projects.

also called for a study (over the same time frame) of "the feasibility and necessity of decontaminating dredged materials [from Baltimore Harbor] and the feasibility of dewatering and recycling [these] dredged materials for use as marketable products"—including examination of "requirements and locations for a processing or staging area, . . . the marketability of potential products, and . . . financial costs." In addition, Section 334[94] authorized a USACE study "on environmentally beneficial ways to expand or supplement existing placement options and sites serving channel dredging operations of the Port of Baltimore." This study was designed to "enhance an ongoing [state-federal] long-term management study for the Chesapeake Bay area. . . ." The report to Congress (due by April 1, 1994) was to discuss results, including (1) demonstrated "beneficial uses of dredged materials to enhance public recreational opportunities, increase living resource habitats, and enhance the environmental quality of Chesapeake Bay"; (2) identified "areas for beneficial use placement of dredged materials to enable the Port of Baltimore to continue maintenance dredging until a long-term management study recommends viable alternatives"; and (3) developed "options for beneficial use placement of dredged materials for each site identified. . . ."[95]

Section 316[96] mandates a study of "the feasibility of establishing a transfer facility at the Leonard Ranch property owned by the Sonoma Land Trust and adjacent to Port Sonoma, Marin, California, for the drying and rehandling of dredged material from San Francisco Bay which is to be transported to a land site for beneficial uses, including lining, capping, and cover material for sanitary landfills, levee maintenance, and restoration of subsided agricultural lands."

Section 326(e)[97] mandates a study "to identify appropriate remediation techniques (including isolation and treatment) for mitigating dioxin contaminated sediments at their sources." The intent of this provision is "to reduce the problems associated with the dredging and disposal of dioxin contaminated sediments" in New York Harbor and the New York Bight without encumbering or delaying scheduled dredging projects. In addition, Section 405[98] directed the USACE and EPA, using decontamination technologies identified under WRDA 1990 Section 412(c), to "jointly select removal, pre-treatment, post-treatment, and decontamination technologies for contaminated marine sediments for a decontamination project in the New York/New Jersey Harbor." The EPA and the USACE were, after the selection of technologies, to recommend jointly "a program of selected technologies to assess their effectiveness in rendering sediments acceptable for unrestricted ocean disposal or beneficial reuse, or both." "Decontamination" is

---

[94] 106 Stat. 4852–4853.
[95] This study, like many others under WRDA and other statutes, was an "unfunded mandate" that was never carried out because of a lack of funds.
[96] 106 Stat. 4847.
[97] 106 Stat. 4850–4851.
[98] 106 Stat. 4863.

defined broadly to include "local or remote prototype or production and laboratory decontamination technologies, sediment pre-treatment and post-treatment processes, and siting, economic, or other measures necessary to develop a matrix for selection of interim prototype[s] of long-term processes." A technique "need not be preproven in terms of likely success."

Section 327[99] directs the USACE to "conduct a national study on information that is currently available on contaminated sediments of the surface waters of the United States" and to compile the resulting information "for the purpose of identifying the location and nature of contaminated sediments in the Nation."[100] The resulting report to Congress (due by April 1, 1993) was to include "recommendations for the collection of additional data on the contaminated sediments. . . ." Section 328[101] authorized the USACE to cooperate with nonfederal interests (to the extent of $200,000 per year) "in the completion of a study of contaminated sediments in Milwaukee Harbor, Wisconsin, and surrounding areas."

Section 345[102] required a study on "bank stabilization and marsh creation by construction of a system of retaining dikes and by beneficial use of dredged material along the Calcasieu River Ship Canal, Louisiana, at critical locations." The report was to include recommendations for specific beneficial use measures.

Section 356[103] directed the USACE (by November 1, 1993), in coordination with the Toledo Port Authority and the Ohio Environmental Protection Agency, to "develop a comprehensive five-year and 20-year sediment management strategy for the Maumee River, Toledo Harbor." This strategy "may include a combination of several sediment disposal alternatives and shall emphasize innovative, environmentally benign alternatives, including reuse and recycling for wetland restoration."

STATE REGULATORY PROGRAMS

The regulatory context for managing contaminated marine sediments and the potential array of management responses and technology choices is bound up inextricably with the fundamental definition of a "contaminated" marine sediment and the distinction between contaminated and "clean" marine sediments. The Congress has expressed a strong preference in this area for respecting local standards as reflected through two relevant programs: the Section 401 water quality certification process under the CWA and the coastal zone consistency review process through the CZMA.

---

[99] 106 Stat. 4851.

[100] This data collection requirement tracks one of the recommendations of the 1989 NRC study on contaminated sediments (NRC. 1989. *Contaminated Marine Sediments: Assessment and Remediation.* Washington, D.C.: National Academy Press).

[101] 106 Stat. 4851.

[102] 106 Stat. 4858.

[103] 106 Stat. 4860.

Section 401 Water Quality Certification

Section 401(a) of the CWA states that any applicant for a federal license or permit for any activity that may result in a discharge of pollutants to navigable waters must provide certification of compliance with the standards and limitations of the state having jurisdiction at the point of discharge. Although the obvious application of this section is to permits issued under the authority of the CWA, it may also apply to ocean dumping permits issued under the MPRSA (at least for dredged material dumped within a state's territorial waters).[104]

Under the CWA, states are authorized to establish water quality standards for waters within their jurisdiction.[105] The geographical reach of that jurisdiction depends on how each state defines the limits of its waters. At the recommendation of the EPA, many states are in the process of rewriting their water quality regulations to include wetlands in the definition of state waters and to establish different standards for wetlands and free-flowing waters. As the definitions of state waters become broader, so does the scope of the resource areas that must be considered in evaluating the compliance of activities contemplated by permits issued under the CWA and MPRSA.

By definition, activities that need a CWA Section 404 permit contemplate a discharge of pollutants into navigable waters, and thus they always require state certification. Section 401 gives states the authority to prevent any federal or private action from proceeding until compliance with state water quality standards can be demonstrated by the permittee or federal agency, including the USACE. Certifying states have the power either to certify permit compliance, to certify with conditions, or to deny certification. If certification is denied, then the USACE cannot issue the permit. A state may take up to a year to make its determination.[106] The water quality certification decisions of a state are subject to judicial review in the courts of the certifying state as a matter of state law.

---

[104] As applied to ocean dumping, MPRSA is not clear on the states' jurisdictional status regarding CWA Section 401 water quality certification. Although Section 401 does apply to all federal permits, only those states in which the discharge originates may certify compliance. Otherwise, states are relegated to the role of "affected states" pursuant to CWA Section 401(a)(2), which gives them a role, but they do not have the final say. However, the conditional language of Section 401, which grants certifying jurisdiction to a state in which discharge *may* occur, suggests that discharge in the state's territorial waters incidental to ocean dumping may require state certification. This conclusion is supported by the reasoning in *Save Our Fisheries* v. *Callaway,* 387 F. Supp. 292 (D.R.I. 1974), in which the court found that state certification of an ocean dumping permit was not required. The court did not find that the state lacked jurisdiction to review certification but that the activity was exempt from review as a federal agency action. The court implied that state water quality certification would have been necessary if a federal agency had not been the acting party (387 F.Supp. at 306). Of course, the exemption no longer exists for federal agencies, leaving both federal actions and private party permit applicants open to state water quality review.

[105] 33 USC §1313(a).

[106] 33 USC §1341(a).

Section 510 of the CWA allows states to adopt water quality standards that are more stringent than those adopted by the EPA.[107] These two state powers—certification review and water quality standards promulgation—grant the states considerable power to control activities that otherwise would be in the domain of the federal government.[108]

More recent laws have shown that the federal-state relationship with respect to this issue continues to evolve. Under MPRSA Section 102(a), no permit may be issued for the dumping of material that will violate "applicable [state] water quality standards." Under WRDA 1992, states are given plenary authority to "adopt or enforce any requirement respecting dumping of materials into ocean waters within the jurisdiction of the State." Although this is subject to certain limitations in the case of federal projects, WRDA 1992 is one indication that the states' jurisdictional status is still evolving.

USACE general regulatory policies also incorporate the requirement that the water quality of affected states be considered. They state that "[a]pplications for permits for activities which may adversely affect the quality of waters of the United States will be evaluated for compliance with applicable effluent limitations and water quality standards. . . ."[109] In its ocean dumping regulations, the EPA also requires that permit applications consider the potential impacts on applicable water quality standards.[110] This requirement is based on the EPA's interpretation of its statutory mandate that ocean dumping permit application review must consider the impact of the proposed dumping on "human health and welfare, including economic, aesthetic, and recreational values."[111]

In some instances, a state has refused without explanation to accept the USACE (or a port's) proffered demonstration that a proposed dredged material discharge will not violate applicable water quality standards, declining to either grant or deny the requested water quality certification under CWA Section 401. (Sometimes a state simply requests additional information, without stating why the existing submittal was considered insufficient.) Where a project may be politically unpopular, these techniques can be used by a state as devices for effectively vetoing the project without ever having to articulate any water quality rationale or other technical justification. An example of this was the withholding of

---

[107] 33 USC §1370.

[108] Indeed, CWA Section 401 originally did not apply to federal agency action. Section 401(a)(6) formerly provided that "[n]o Federal agency shall be deemed to be an applicant for the purposes of this subsection." This provision was held by the Rhode Island Federal District Court to insulate the USACE and its private contractors from the application of state water quality standards under the CWA (*Save Our Fisheries v. Callaway*, 387 F. Supp. 292 [D.R.I. 1974]). However, the 1977 amendments to the CWA removed and replaced Section 401(a)(6), eliminating the federal agency exception.

[109] 33 CFR §320.4(d).

[110] 40 CFR §227.18(b–c).

[111] 33 USC §1412(a)(B).

water quality certification by New York state for a proposal to deposit dredged materials into excavated subaqueous pits, followed by capping with clean materials. Regardless of whether this was a case of a state arbitrarily or unjustifiably withholding certification or a matter of the USACE refusing to comply with a legitimate request for additional information, procedural disputes of this type are costly and inefficient. There need to be agreed-on procedures for demonstrating compliance with water quality standards; there needs to be a clearly specified, reasonable time limit for obtaining a "yes" or "no" decision; and certification denials need to have clearly articulated, technically supportable rationales related to water quality.

Coastal Zone Consistency Review

Section 307(c) of the CZMA was established in 1972 to require, among other things, that federally conducted and regulated activities comply to the "maximum extent practicable" with states' federally approved coastal zone management plans. Any coastal sediments management activity regulated by the USACE or other federal agency must provide a state determination of consistency with the state's federally approved coastal zone plan. Section 307(c) applies to all coastal states and coastal Great Lakes states. Section 307(c) has evolved into a program that regulates activities through the consistency determination process. This NOAA-administered program can delay sediment management activities.

The federal agency or applicant provides a determination of consistency to the state coastal zone management agency. The state agency either concurs or objects. In the case of objections, the state must tell the regulated entity what must be done to bring the project into consistency.[112]

States have a great degree of latitude in interpreting and administering their coastal zone plans. State coastal zone plans are, by their very nature, general and often vague with few, if any, exact requirements or standards that a federal agency or applicant might use as a basis for determining compliance. As such, the Section 307(c) compliance process is frustrated by state requirements, conditions, and controls that are often difficult to accomplish, procedurally cumbersome, and outside the purview of the federal agency or applicant to accomplish. Unlike proposed development activities for which the consistency determination process can be used as a guide, existing contaminated sediments are already in place, and the CZMA provides no authority for removing the sediments or regulating their management through standards.

Coastal zone plans cover very broad jurisdictional boundaries. The Mississippi

---

[112] In the case of a federally permitted project, state concurrence is required before a permit can be issued. In the case of a project directly undertaken by a federal agency, it is up to the agency to "ensure . . . to the maximum extent practicable" that the project is consistent with the state's enforceable coastal zone management policies.

coastal zone encompasses the three coastal counties, regardless of elevation, whereas Alabama uses a contour 15 feet above mean sea level. Florida has determined that the entire state is in the coastal zone.

The federal coastal zone consistency process does not encourage conclusion of review by the state. Thus, the opportunity for delay exists when a state might not support a federal project as part of federally regulated activity. This is illustrated in the proposed Texas coastal plan that directs beneficial use of all dredged material. Endless negotiations by the USACE will take place while attempting to obtain congressional funding to comply with this provision. Managing contaminated sediments while attempting to comply with state coastal zone plans can be perilous, to say the least.

States are not limited in their review. States often perceive that consistency determinations are required for an entire project, when in fact only one segment may be under review by the federal agency. This perception changes the influence of the state over the projects under consideration.

Most often, state consistency requirements are regulatory in nature but are only tied to a broad coastal zone plan. For example, the USACE often is required to obtain CWA Section 401 water quality certification for a dredged material disposal plan. The connection between compliance with numeric state water quality standards and a state coastal zone plan is remote at best. Moreover, without standards for compliance, wide latitude in interpretation can frustrate compliance for sediment management activities from year to year and with each change of state administrations.

Other State Jurisdictional Issues

Another area in which states exercise local authority over federal agency action and permitting decisions is where an activity causes a seaward extension of the regular low tide mark. The division of ownership of underwater lands was resolved generally by the Submerged Lands Act,[113] which was enacted in 1953. The Submerged Lands Act provided that, barring other particular claims, states had the rights to submerged lands up to three miles from the regular low tide mark, and the federal government held title to submerged lands beyond that point.[114]

In 1993, however, the U.S. Supreme Court considered the situation in which a state navigation project would cause an accretion of the coastline.[115] Under the terms of the Submerged Lands Act, such an extension of the coastline would entail a concomitant extension of the state's title to submerged land. Prior to

---

[113] 43 USC §1301 *et seq.*
[114] 43 USC §1301(b).
[115] *United States v. Alaska*, U.S. 112 S.Ct. 1606 (1992) (unanimous decision).

issuing a permit to conduct the necessary dredging and filling work, the USACE submitted the permit application to the DOI which objected to the project and recommended that the USACE require the state to waive its right to an extension of its submerged lands.[116] The Supreme Court upheld this requirement as a condition to the issuance of the permit, and generally held that in similar situations it was necessary for the federal government to protect its property interests.[117]

What the court did not consider, but is suggested by implication, is that a state could block a project causing such a coastal accretion by refusing to grant a waiver to its claims of title to additional submerged lands. This was not a likely problem in *United States* v. *Alaska* because the project was one pursued by the state itself. In other circumstances, however, it is not unreasonable to suggest that a state could assert its right to additional submerged lands when other means of project opposition, such as water quality certification, have proved unsuccessful.

Another area in which the USACE has relinquished some degree of its authority, and in which its actions and options regarding the management of contaminated marine sediments may be influenced, concerns the applicability of local zoning regulations. The USACE general regulatory policies provide that the primary responsibility for determining zoning and other land use matters rests with the state and local governments and that the USACE generally will accept such decisions. The USACE maintains the authority, however, to ignore local decisions when it finds issues of "overriding national importance," which may include "national security, navigation, national economic development, water quality, preservation of special aquatic areas, including wetlands, with significant interstate importance, and national energy needs."[118]

When a Section 404 permit application pending before the USACE is denied by local authorities pursuant to zoning regulations, the USACE will consider that denial, and depending on the stage of its own decision on the permit, will take one of three actions: immediately deny the application as against the public interest, deny the application without prejudice to be renewed, or approve the permit application notwithstanding the local zoning conflict.

This final option is only available when the USACE makes a determination that the national interest is at stake, and the USACE permit acts to override a local denial. Without such a determination, the permit applicant must redouble efforts at the local level to gain local permit approval.

One of the most viable beneficial uses applicable to contaminated marine sediments is extending the shoreline as part of near-shore CDFs (as illustrated in the Port of Tacoma project—see Appendix C) or constructing or restoring

---

[116] 112 S.Ct. at 1609.
[117] 112 S.Ct. at 1615–1619.
[118] 33 CFR §320.4(j)(2).

offshore islands (as illustrated in the Hart and Miller islands project—see also Appendix C). Such use allows contaminated sediments to be isolated and contained by placement in the interior of the diked area, surrounded and covered by progressively cleaner materials, while still taking advantage of their physical bulk. The legal issue of who owns the real estate thereby created can be significant, particularly in land-scarce urban areas, where such real estate can be very valuable. The legal complexities of this issue are illustrated by the dispute, now pending in the U.S. Supreme Court, between the states of New York and New Jersey over who owns Ellis Island in New York Harbor. Although the original island was in New York waters, the island was expanded by the use of fill, so that a large part of the island is now on the New Jersey side of the line separating the two states' waters. Where dredged material is used to construct or expand an island, a case may become even more complex, depending on who owned the dredged material used for construction.

GAPS, OVERLAPS, AND UNCERTAINTIES

Table B-1 indicates how a few federal statutes, and potential state approval requirements, may apply to six sediment excavation and management scenarios. Table B-1 illustrates the complexity of the regulatory framework, showing that multiple legal authorities may apply simultaneously in a given situation (e.g., Scenario 4), whereas in other cases (e.g., Scenario 6) there is the possibility that no statute applies. Table B-1 also contains many footnotes, reflecting the confusion and uncertainty over the applicability of certain statutes in particular situations. Readers confronting such situations are encouraged to consult knowledgeable environmental counsel. As can be seen, a Section 10 RHA permit would be required any time excavation or dredging is carried out in navigable waters (e.g., scenarios 1 and 6)[119] and could be required whenever construction or capping is carried out in navigable waters, including coastal ocean waters (e.g., scenarios 2 and 4).[120]

In addition, by virtue of the "excavation rule" promulgated in August 1993,[121] any excavation, mechanized land clearing, or channelization work in waters of the United States presumptively requires a CWA Section 404 permit (e.g.,

---

[119] In the context of the RHA, "navigable waters" refers to waters that are "navigable-in-fact" in the traditional sense or susceptible to navigation. The term does not apply to more broadly defined "waters of the United States," including wetlands, which may be used by migratory birds or have some other link to interstate commerce but are not susceptible to navigation.

[120] Whether Section 10 is considered to apply or not depends on whether there is a demonstrable obstruction to navigation or the potential to alter the course, location, condition, or capacity of a navigable waterway. In addition, Section 10 might not apply where a construction project in navigable waters was authorized specifically by Congress.

[121] Federal Register, vol. 58, no. 163, August 25, 1993, pp. 45007–45033; 33 CFR §323 and §328.

TABLE B-1 Interrelationships of Sediment Regulatory Authorities in Selected Scenarios

Scenario	A. CWA	B. MPRSA/LDC	C. RHA	D. CERCLA	E. RCRA	F. State approvals
1. Excavation of contaminated sediment hot spots from a waterway	Section 404: Permit required for excavation in 404 waters (including wetlands, if not part of normal dredging operations[a])	Not applicable	Section 10: covers dredging and excavation in navigation channels regardless of purpose (i.e., navigational dredging versus environmental cleanup)	Could apply when part of the cleanup or restoration of an underwater Superfund site	If part of a corrective action or RCRA facility closure	State CZMA consistency determination could be required if conducted in state coastal zone; state certification under CWA 401 could also be required for activities requiring a CWA permit (see 1.A)
2. Use of sediments to construct berms, containment facilities, or islands in navigable or ocean waters	Section 404: Permit required for discharges of dredged or fill materials in 404 waters[b]	Section 103: Permit required for disposition of any material in ocean waters (seaward of the baseline); no permit required when material is placed to construct an artificial island when otherwise regulated by federal or state law	Section 10: Permit required for obstructions to navigation and changes in the course or condition of navigable waters	Not applicable[c]	Not applicable[d]	See 1.F
3. Ocean dumping of dredged material	Section 401: State water quality certification could be required for dumping in or near state territorial waters, when the dumping may cause state water quality standards in such waters to be exceeded[e]	Section 103: Permit required from USACE for the transportation and dumping of this material	See 2.C	Not applicable (but see footnote b)	Not applicable (but see footnote c—at least for material dumped beyond the three-mile limit of the territorial sea)	Section 401: State certification could be required if discharged in the territorial seas; CZMA consistency determination could be required for loading and transport facilities located in a state's coastal zone

continues on next page

TABLE B-1 Continued

Scenario	A. CWA	B. MPRSA/LDC	C. RHA	D. CERCLA	E. RCRA	F. State approvals
4. Disposal of contaminated sediments followed by clean capping	Section 404: Permit required if in 404 waters (or involving a return flow to such waters)[f]	Section 103: Permit required if discharged into ocean waters[g]	Section 10: Permit may be required if intentional capping results in mounding of sediments that obstructs navigation or alters the waterway's condition	If contaminants are subsequently released and cause an environmental hazard, cleanup could be required, unless specifically permitted under 4.A or 4.B; capping may or may not be deemed a preferred remedy under Superfund	If sediments are TC toxic (exceed RCRA TCLP limits), the discharge occurs in inland or near-shore waters (out to three miles)—especially wetlands, and the disposal is not permitted under 4.A or 4.B, a RCRA permit could be required	See 3F.
5. Land disposal of contaminated sediments	Runoff or return flow into 404 waters may require a permit under Section 404 or Section 402[h]; a 404 permit would also be needed if the disposal site includes a regulated wetland	Not applicable	Not applicable	If land disposal subsequently results in uncontrolled contaminant releases (especially if they cause an imminent hazard, cleanup or natural resource damage restoration could be required under CERCLA—unless the specific contaminants were authorized under a federal permit	The USACE has asserted since 1988 that RCRA does not apply to any land disposal of dredged material, but the EPA does not agree; a pending RCRA rule would exempt from the possibility of RCRA regulation for on-land dredged material containment facilities that have 404-regulated return flow or that impact regulated wetlands (see also 4.E)	If 5.A applies, section 401 certification may be required; if located in a state's costal zone, a CZMA consistency determination may be required (see 1.F)

217

6. Natural restoration	Not applicable	Not applicable	Studies could be required under Superfund (both remedial action and damage restoration provisions), leading to selection of the natural restoration alternative	If site is considered a RCRA site (see 4.E), natural restoration may or may not be considered adequate corrective action	When state has assumed the RCRA program under 4.E situation; otherwise, not applicable

[a]This became a regulatory requirement under the excavation rule adopted by the USACE and EPA in the summer of 1993. A 404 permit would thus be required, covering the incidental fallback of sediment, when sediments were excavated for environmental cleanup or as a source of fill material.

[b]Although there may be some debate on this point, sediments dredged from a waterway bottom are dredged materials even if they are used as fill to convert an aquatic area to dry land. The CWA regulates dredged material discharged in inland waters out to the coastal baseline. Beyond this point, the MPRSA takes over. Fill material that does not come from a waterway bottom is not regulated under the MPRSA (unless it qualifies as the disposal of a waste) but requires a 404 permit when deposited out to the three-mile limit of the territorial seas.

[c]Unless the material is contaminated and results in an uncontrolled release requiring remediation that is not specifically addressed in a federal permit (and thereby subject to the federally permitted release exemption from CERCLA).

[d]If the facility is considered to contain hazardous wastes or to be a hazardous waste treatment, storage, or disposal facility, the EPA *could* consider it to require a RCRA permit. However, pending revisions to RCRA regulations would create a dredged material exclusion for sediments that are otherwise regulated by permit under the CWA or the MPRSA. This would include confined disposal facilities with a return flow to U.S.-regulated waters, requiring CWA Section 404 permit..

[e]See Title V of WRDA 92 and discussion in the text.

[f]A permit will be issued or withheld, depending on the extent of the biological, physical, and chemical impacts to the aquatic environment. The mitigating effects of capping can be considered.

[g]A permit will be issued or withheld, depending on the results of bioassay and bioaccumulation tests and other evaluations required under the ocean dumping criteria. If the material is unsuitable for ocean dumping, it is unclear legally whether after-the-fact capping can be used to render permissible an otherwise prohibited activity. Parties to the London Convention of 1972 have signified their intent under the convention that management practices, including capping, that reduce impacts on the marine environment can be taken into account in deciding whether materials may be disposed of at sea. The EPA is considering amendments to the ocean dumping criteria to clarify that this is also the case under U.S. law.

[h]A Section 402 effluent discharge permit could be required if the land disposal site includes treatment of runoff or wastewater and the effluent is discharged from a point source (e.g., an outfall pipe).

scenarios 1 and 6). The only exception was for navigation dredging in traditionally navigable waters.[122] The excavation of contaminated sediment hot spots for environmental reasons would not qualify for this exception.[123]

A state water quality certification under CWA Section 401 is not required for dredging except when covered by the excavation rule (see Scenario 1.A), but it would be required whenever a discharge would have the potential to adversely affect the quality of waters of the United States subject to state jurisdiction (i.e., out to the three-mile limit of the territorial seas). Any activity (including indirect staging, transporting, and handling) in or affecting a state's "coastal zone" would be subject to a determination of consistency with a state's federally approved coastal zone management plan (see scenarios 1, 3, and 6).

Dredged or excavated uncontaminated sediments can be dumped only at officially sanctioned, formally designated ocean dump sites—and then, only subject to a USACE Section 103 permit under the MPRSA. Contaminated dredged materials that fail prescribed bioassay and bioaccumulation tests under the ocean dumping criteria and Green Book guidance (or that are otherwise deemed to violate prohibitions under the London Convention of 1972 against the ocean dumping of "wastes and other matter" containing "Annex I" constituents as "other than trace contaminants") could be barred from ocean dumping (or could be subject to stringent management controls, including capping)[124] (see scenarios 2, 3, and 4). Under WRDA 1992 Title V (which amended the MPRSA), in addition to enforcing state water quality standards (through the 401 certification process), states now have the authority to establish their own restrictions on ocean dumping in waters subject to state jurisdiction.

A CWA Section 404 permit is required from the USACE—again, subject to state water quality certification—when excavated sediments are to be placed (or capped) in inland waters (i.e., landward of the coastal baseline) or wetlands (see scenarios 2 and 4). If dredged sediments from the seabed or inland waters were

---

[122] The excavation rule has been challenged by several industry trade associations in American Mining Congress, et al. v. U.S. Army Corps of Engineers, et al., Civil Action No. 93-1754 (D.D.C. 1993).

[123] However, if the excavation were carried out pursuant to a CERCLA record of decision as part of an approved remedial action, then the need for a formal permit could be avoided as long as the underlying substantive requirements of the Section 10 regulations were satisfied.

[124] Although the London Convention purports to flatly prohibit the ocean dumping of wastes or other matter containing Annex I ("black list") constituents as "other than trace contaminants," the United States employs bioassay and bioaccumulation—rather than chemical—tests to determine whether this situation exists. Because such tests are not chemical specific (i.e., one cannot be certain whether an Annex I constituent or something else caused the test to be failed), and because there are escape clauses that have been recognized under both international and domestic law, the prohibition is not quite as absolute as its wording would suggest. As previously noted, the ocean dumping criteria contain exceptions (e.g., for material that is "rapidly rendered harmless") that could allow material to be dumped in the ocean even after the biological tests have been failed.

discharged[125] intentionally by pipeline into ocean waters out to three miles, the sediments likewise would be subject to the CWA (i.e., a Section 404 permit—and possibly an ocean discharge permit under Section 403) rather than to MPRSA requirements. The same (i.e., the need for a Section 404 permit) holds for "runoff or overflow [into 404 waters] from a contained land or water [dredged material] disposal area," which USACE regulations[126] define as included in the definition of "discharge of dredged material" (see scenario 5.A).

If contaminated sediments are excavated for environmental cleanup purposes, any disposal or management action (including capping) could be subject to the remedial cleanup requirements of CERCLA—if there is an uncontrolled release triggering the need for such cleanup (see scenarios 1.D, 4.D, and 5.D). Indeed, as part of the development of a remedial action plan, even natural restoration (see scenario 6.D) could require regulatory review and approval.

Moreover, if contaminated sediments are placed on land or at a coastal disposal site or containment facility, RCRA could be deemed to apply—if the sediments display hazardous waste "characteristics."[127] The USACE long has taken the position that dredged material is exempt from RCRA because it is not "solid waste." The EPA's view is that it can still be subject to RCRA (under the "contained-in rule"), when it becomes contaminated with hazardous wastes, pollutants, or contaminants. Under the hazardous waste identification rule under development by the EPA, consideration is being given to a "dredged material exclusion," which would exclude dredged materials (but not "fill material") from the possibility of RCRA regulation *if* they are being regulated under a CWA or MPRSA permit. This would include dredged material placed in an on-land CDF with a regulated "return flow" but not dredged material in a totally contained CDF (see scenarios 4.E and 5.E).

Finally, in cases where in-place contaminated sediments cause or contribute to injury or the loss of "natural resources," including coastal biota, parties responsible for the contaminant releases can be sued by federal, state, or tribal "trustees" for natural resource damages. Recovered money is used not only to pay for cleanup of the affected sediments, but also to fully restore, rehabilitate, or provide the equivalent of the injured resources. Such monetary damages can be substantial because they include compensation for any interim "lost use" of the

---

[125] USACE rules are based on intent.

[126] 33 CFR §323 and §323.2(d).

[127] The most likely "characteristic" to apply in this situation is the "toxicity" characteristic, the presence of which is established by subjecting the material to the TCLP and determining whether established regulatory levels for any of 39 specified TC constituents have been exceeded (40 CFR §261). Also, see *Federal Register*, vol. 55, no. 126, June 29, 1990, pp. 26986–26998. Wet dredged material is unlikely to exceed TC regulatory limits, even when high levels of contaminants are present. Dry dredged material could exceed such limits when certain contaminants are present, even at very low levels.

injured resources and even for the lost "contingent value" of knowing that the resource is there in an unimpaired condition (see scenarios 1.D and 5.D).

Gaps in Coverage

The preceding discussion demonstrates that there are a number of overlaps in regulatory coverage of sediments that are managed because of requirements associated with either navigation dredging or environmental cleanup. It also provides a flavor of the complexity associated with a hodgepodge of laws and regulations that were never primarily intended to address sediments. And it provides an indication of some of the uncertainties related to the applicability of particular regulatory requirements under certain circumstances.

There are also gaps in coverage. Consider the dilemma of a coastal port that must dredge its harbor and navigation channels periodically. These waters are subject to contamination from a multiplicity of point and nonpoint sources, as well as from spills, both in the immediate port area and throughout wide areas of tributary watersheds. In most cases, the vast majority of contaminants that come to rest in bottom sediments are derived from sources located many miles upstream. Yet the port bears the brunt of the impact—in terms of increased placement costs for contaminated sediments and in terms of lengthy delays in securing the necessary regulatory permits and approvals—if they can be secured at all. By contrast, the responsible upstream sources are seldom held accountable. Occasionally, they may have their discharge permit limits tightened. Infrequently, they may be subject to cleanup requirements or restoration orders under CERCLA. But there is no systematic mechanism in place for ensuring that upstream discharge sources take into account downstream impacts before being allowed to discharge[128] or, failing this, that responsible upstream sources share in the incremental costs they impose on downstream users, such as ports.

Another illustration of legal uncertainties relates to the placement of contaminated sediments within a diked containment area for the purpose of constructing an offshore containment island. The status of this material, and even the legality of its placement, depends on whether it is deposited more than three miles offshore (see Table B-1), whether the "purpose" or "primary purpose" of placement is considered disposal or island creation, and/or whether an exclusion under the ocean dumping criteria applies (see Scenario 2.B).

The MPRSA (33 USC §1402[f]) defines "dumping" to exclude ". . . the construction of any fixed structure or artificial island [and] the intentional placement of any device in ocean waters or on or in the submerged land beneath such waters, for a purpose other than disposal, when such construction of such

---

[128] Limited exceptions to this are the water-quality-based discharge limits under Section 303 of the CWA, which were discussed previously.

placement is otherwise regulated by Federal or State law or occurs pursuant to an authorized Federal or State program. . . ." Clearly, if clean sand of land origin were deposited to construct a containment dike within which an island would be created, the placement of this "fill material" would be exempt from the MPRSA—both as "fill" (if carried out within three miles, where it would be regulated under CWA Section 404—see scenario 2.A) and under the MPRSA's exclusion for construction of a fixed structure. For contaminated sediments (i.e., of aquatic origin), however, it would be harder to argue both that the material was not MPRSA-regulated "dredged material" and that the material was being deposited "for a purpose other than disposal." At best, one could say that, in addition to disposal, one purpose is the creation of an offshore island. It is not clear whether this is sufficient. Also, if the material were deposited in ocean waters beyond the three-mile limit, where no other federal or state permit would be required, it is unclear whether it could qualify under the MPRSA's "construction" exception, which requires regulation under other federal or state law.

Ultimately, if the sediment contaminants were isolated sufficiently—i.e., by being encapsulated within a containment dike and surrounded by large volumes of uncontaminated sediments—from the surrounding marine environment so as to be rendered "non-toxic to marine and non-bioaccumulative," then the material might be viewed, despite its intrinsic toxicity, as present as "other than trace contaminants" under the exception found in the ocean dumping criteria.[129] Alternatively, one might argue that confined disposal on the seabed behind containment dikes is "seabed emplacement" and not "disposal in ocean waters" and that it is appropriate to treat it as similar to land containment in a CDF (see discussion of related issue in Table B-1).

Although the use of contaminated sediments to construct offshore containment islands probably will become more common, in the New York Bight and elsewhere, as open-water and land options become more limited, resolution of this legal issue may have little practical significance because a project of this magnitude almost certainly would require separate congressional authorization. And Congress is free to specify what permitting and environmental analysis procedures it wishes to have applied.

POTENTIAL REGULATORY REFORMS

The following legislative and regulatory changes could be considered to avoid or minimize gaps, overlaps, uncertainties, and inefficiencies:

Amendments to the CWA could distinguish between the open-water disposal of dredged material and the discharge of fill material into wetlands and other "special aquatic sites." Much of the complexity that has developed over the years

---

[129] 40 CFR §227.6(f).

in the application of Section 404 is the result of attempts to control adverse effects on wetlands caused by development. A new CWA section dealing with open-water discharges of dredged material would allow the separation of wetlands development and protection issues from the very different issues associated with the construction and maintenance of efficient navigation channels.

The open-water disposal of dredged material resulting from navigation dredging of public ports and waterways could be subject to a unified set of statutory and regulatory requirements that do not differentiate (except where a specific technical justification can be provided) among inland, estuarine, and ocean waters. The EPA and the USACE have gone part of the way toward accomplishing this by developing an Inland Testing Manual, which parallels the Green Book in setting forth a tiered testing framework using freshwater and estuarine species.

The decision-making framework for the management of contaminated sediments could be simplified and made more efficient if the full range of placement and management options were required to be considered, based on environmental acceptability (risk), technological feasibility, and economic viability. From the standpoint of environmental acceptability and risk, a single, unified tiered-testing procedure needs to be established and used to define the contamination status and environmental and health hazard potential of sediments excavated from navigable or ocean waters. This procedure could address the environmental risks associated with freshwater, marine, and land placement, containment, or the beneficial use of such sediments.[130]

The screening criteria and regulatory limits associated with this procedure could be treated as ARARs under CERCLA and could preempt (by statute) otherwise applicable federal regulatory requirements under the RCRA and other pollution control statutes. (This procedure should not, however, displace CWA Section 404 wetland procedures or other requirements directed primarily at preventing direct physical damage or disturbance, rather than pollution, impacts.)

In-place or off-site capping of contaminated sediments, where determined by the EPA or USACE to be an environmentally acceptable, economically viable, and technologically preferred alternative, could be deemed to be a form of "treatment [which] permanently and significantly reduces the . . . toxicity or mobility" of associated pollutants and contaminants, within the meaning of Section 121(b) of CERCLA.[131]

Because there can be conflict between the state and federal requirements that are difficult to resolve, the EPA could be given the statutory authority to reject as scientifically unjustified more-stringent state requirements—unless the state is

---

[130] The EPA is standardizing solid-phase sediment toxicity and bioaccumulation tests for freshwater and marine species. See U.S. Environmental Protection Agency. 1992. *Tiered Testing Issues for Freshwater and Marine Sediments*, EPA 823-R93-001. Washington D.C.: EPA.

[131] 42 USC §9621(b).

APPENDIX B   223

prepared to provide an alternative site or assume the incremental costs of additional placement or use restrictions. This change might foster timely decision making.

The "local cooperation" requirement (e.g., under WRDA 1986 Sections 101, 102), which currently requires local sponsors of federal navigation projects to bear full responsibility for the construction, operation, and maintenance of necessary dredged material disposal sites, could be repealed or substantially modified. This is needed to avoid the strong economic incentive favoring the open-water disposal of (even highly contaminated) dredged materials (i.e., at "free" disposal sites) in preference to containment on land (i.e., at sites that must be paid for, if not fully, at least in part by the local sponsors). All versions of the proposed WRDA 1996 legislation would require federal cost sharing for on-land dredged material placement sites, including CDFs, so this suggestion is expected to become law.

The ownership status of newly created offshore containment islands, nearshore containment areas, and other new real estate created with dredged material needs to be clarified—that is, ports need to acquire an ownership interest commensurate with their degree of cost sharing or funding of the site and their contribution of dredged material to construction of the new real estate.

The decision-making framework needs to encourage and promote appropriate beneficial uses of dredged material and contaminated sediments.[132] For example, contaminated sediments could be used safely and beneficially in the interior of a diked containment facility, where they are surrounded and capped by uncontaminated sediments.

Legislative initiatives to require watershed planning and management initiatives to control water pollution sources need to take into consideration ports' interests in minimizing upstream point- and nonpoint-source contributions to downstream contamination and need to require explicit consideration of downstream impacts. They also need to require watershed-specific inventories (including identification of sources) of upstream pollutant contributions to problematic downstream sediment contamination in port areas. Federal and state regulatory agencies need to be authorized and required specifically to tighten upstream discharge permit conditions and restrictions to reduce downstream port impacts. Finally, consideration needs to be given to authorizing the EPA, where identifiable upstream sources contribute "disproportionately" to downstream sediment contamination, to allocate and recover an appropriate share of cleanup or disposal costs from such discharge sources.

CWA Section 303 could be amended to require states and the EPA, in setting

---

[132] For example, see U.S. Environmental Protection Agency. 1992. Evaluating Environmental Effects of Dredged Material Management Alternatives—A Technical *Framework*, EPA 842-B-92-008. Washington D.C.: EPA.

TMDLs for waterway segments and in developing load allocations (WLAs and LAs) for point and nonpoint sources, to consider impacts on downstream sediment quality—where such impacts may impair downstream water uses or interfere with or complicate navigational or environmental dredging. Although it is probably not possible for most sediment contaminants to quantitatively link discharges from individual point and nonpoint sources to site-specific buildups in downstream sediments, there is no reason why presumptive sources of problematic sediment contaminants could not be inventoried and prioritized so that regulators could use the TMDL approach to progressively reduce contamination from the most important sources.

CONCLUSION

The efficient and effective management of contaminated sediments, whether associated with navigation dredging or environmental cleanup, is hampered by both too much and too little legislative and regulatory attention. On the one hand, few aspects of sediment handling, from initial excavation to ultimate disposal, are unregulated. On the other hand, regulatory coverage is haphazard and bears little relationship to underlying environmental or human health hazards and little resemblance to an efficient and coherent process that is predictable or reliable.

As environmental concerns make economically critical navigation dredging more and more difficult, and as the identification, prioritization, and remediation of contaminated sediment sites accelerate, these regulatory limitations will become more evident and constraining. Appropriate legislative and regulatory initiatives could do much to avoid these problems and facilitate the efficient and effective management of all sediments, especially contaminated sediments.

APPENDIX C

Case Histories of Representative Remediation Projects[1]

Frank Bohlen, Peter Shelley, and Kenneth S. Kamlet

To develop an understanding of the factors affecting the management of contaminated sediment sites, the committee reviewed a large number of ongoing and recently completed projects. From this group, six projects, which are considered representative of particular site conditions, regulatory constraints, and classes of contaminants, were selected and examined in more detail. These case histories yielded graphic illustrations of the complexity of the management process and provided a basis for subsequent committee evaluations of remediation strategies and the protocols affecting their implementation.

The criteria used by the committee to select the six projects are shown in Table C-1. The case histories provide examples of coastal, lake, riverine, and estuarine conditions and include both navigation dredging and environmental cleanup (Superfund[2]) projects. A variety of remediation strategies, and both organic and inorganic contaminants, are represented. The lessons learned are largely subjective, rather than formal, rigorously derived conclusions, such as those presented in the text of the report. The lessons were considered throughout the committee's study along with other evidence. The sites selected for evaluation were Boston Harbor, Massachusetts; Hart and Miller islands, Maryland; James River, Virginia; Marathon Battery, New York; Port of Tacoma, Washington; and Waukegan Harbor, Illinois.

---

[1] This appendix has been edited for grammar and style; accuracy is the sole responsibility of the authors.

[2] Superfund is the common name for the Comprehensive Environmental Response, Cleanup, and Liability Act.

TABLE C-1 Selection and Evaluation Criteria for Six Case Histories

			Sites			
Selection Criteria	Boston Harbor	Hart/Miller Islands	James River	Marathon Battery	Port of Tacoma	Waukegan Harbor
Navigation dredging	Yes	Yes	No	No	Yes	No
Hot spot/Superfund	No	No	Yes	Yes	Yes	Yes
Imminent hazard	No	No	Yes	Yes	No	No
Remediation end-point	Water quality standards	Water quality standards	FDA action levels	Cadmium, nickel concentration ≤ 100 ppm	State sediments quality objective (apparent effects threshold)	PCB concentration at 50 ppm
Critical technology selected	On-site capping	CDF	Natural restoration	Chemical stabilizer	Near-shore containment	Thermal desorption
Natural restoration	No	No	Yes	Partial	No	No
Capping	Yes	Yes	No	Partial	Yes (in containment area)	Yes
In situ treatment	No	No	No	No	No	No
Ex situ treatment	No	Limited	No	Yes	No	Yes
Ex situ containment	CDFs; confinement	Containment islands	None	Landfill	Diked containment	Diked containment
Examples of beneficial uses	Landfill cover and cap; real estate	Recreational area, reduced erosion	None	None	Enhanced port and habitat	None
Institutional/intangible factors	Public involvement	Public opposition	Fisheries contamination, high cost of remediation	Wildlife area	Stakeholder cooperation	Human health; Superfund

BOSTON HARBOR, MASSACHUSETTS

The ongoing project in Boston Harbor involves significant navigation dredging, source controls, and the removal and isolation of a large mass of contaminated sediment. Contained aquatic disposal (CAD) is planned.

The harbor supports commercial shipping as well as fisheries and recreational boating. The harbor also has served as a contaminant sink for hundreds of years, receiving municipal and industrial discharges, urban runoff, vessel spills, and flows from several watersheds. Sewage has been discharged in two locations in the harbor for many years.

Background

Litigation in the early 1980s forced the construction of new sewage facilities to allow for diffused discharge 9.5 miles offshore. The improvements are to come on line between 1996 and 2000; interim remedial projects already have made significant improvements in water-column and sediment quality as well as benthic activity. In 1968, the U.S. Congress directed the U.S. Army Corps of Engineers (USACE) to review plans for navigation improvements. The planning eventually led to a 1988 feasibility study for dredging and deepening several channels. The dredged material was to be placed at the Massachusetts Bay Disposal Site (MBDS), some 22 nautical miles east of Boston, in waters 80 to 100 meters deep. Contaminated sediments were to be capped.

The dredging project stalled because of a number of factors, including political changes in the state, growing public concern about the offshore placement of dredged material near an area subsequently designated as a national marine sanctuary, and the tightening of sediment characterization criteria. The USACE and the local project sponsor (the Massachusetts Port Authority or MASSPORT) conducted another environmental review in 1992, this time with extensive public participation. More than 37 organizations (including environmental groups, state water quality certification officers, neighborhood organizations, and businesses) were invited to a meeting, with a facilitator from a professional office of dispute resolution. Two technical working groups were established, one focusing on sediment characterization and one on disposal options. During the process leading to the designation, many more people became aware of and involved in the harbor remediation decision process.

Remediation Alternatives Considered

The screening process began with 312 land-based inland and coastal sites, 21 landfills, and 21 aquatic sites. This list was narrowed to 24, from which four acceptable and preferred alternatives were identified: MBDS, the Boston Lightship site, two near-shore borrow pits, and one CAD site.

The cooperative review led to changes in the decision-making process: (1) the MBDS was dropped from consideration because of uncertainties about the efficacy and durability of capping at the proposed depths and because of persistent concerns about deleterious biological impacts on the nearby marine sanctuary; (2) the use of a cost cap as a screening tool was eliminated, and all alternatives were analyzed primarily on the basis of their technical and environmental merits, with a secondary cost-effectiveness screen; (3) the project team focused on the use of previously impacted areas as placement sites; (4) the USACE agreed to treat all silts as contaminated; and (5) innovative technologies for the handling, treatment, and placement of contaminated marine sediments were evaluated thoroughly. The unit costs of all the alternatives considered ranged from $16/cubic yard ($yd^3$) for open-ocean capping at the MBDS to well over $200/yd^3$ for various treatment or land and shore-side containment facilities.

Strategy Chosen

In the end, an entirely new strategy was chosen: in-channel CAD with sand capping. To accommodate the in-channel CAD strategy, portions of the navigation channels were to be overdredged to form a series of deep cells or pits. Contaminated sediments from the remainder of the channel were to be placed in these pits and capped with a layer of sand at least three feet thick. Clean dredged material from the deeper sediments of each pit was to be barged offshore for placement at the MBDS. It was recognized that in-channel placement would limit future capabilities to deepen the waterway to accommodate vessels of increased draft. This was not considered a major impediment, however, because channel depths already were limited effectively by several roadway tunnels passing under the harbor.

Dredging costs associated with the in-channel disposal option were estimated at $37/yd^3$, resulting in a total project cost of approximately $58 million.

Overall Assessment

It is premature to declare that the project has cleared its last hurdle, but the reactions of the general public, the regional press, and project participants have been positive. Although the project is without question more complicated and expensive than the original proposal, a wider range of stakeholders consider it to be both environmentally preferable and politically implementable.

Lessons Learned

- In a limited number of cases, historically contaminated bottom areas within or close to a dredged navigation channel may represent the optimum location for the disposal or containment of contaminated channel sediments. The use of these areas is not beyond the capability of existing methodology.

- Involvement of all key interest groups at the earliest possible stage in the planning process can contribute significantly to ultimate project acceptability. Dispute resolution tools can be helpful in building consensus.

- When conducting cost-benefit analyses of various management strategies, decision makers need to recognize that considering cost criteria alone is overly restrictive. The most desirable choice appears to be the lowest-cost solution that is *politically and environmentally acceptable*.

HART AND MILLER ISLANDS, MARYLAND

The Hart and Miller islands project involves a diked confined disposal facility (CDF) that was constructed specifically to receive all sediments dredged from Baltimore Harbor and its approach channels within the Patapsco River. By state law, these sediments are considered to be contaminated and must be contained. The site was also intended to receive clean sediments from a major deepening project.

Baltimore Harbor sediments have been contaminated by a variety of agricultural, industrial, and municipal wastes. Bottom sediments dredged from the harbor and its approach channels were to be placed in the 1,100-acre CDF and capped with clean material. The facility is adjacent to Hart and Miller islands, which are located in the Chesapeake Bay several miles north of the harbor entrance. Both of these small islands support abundant wildlife. The capacity of the CDF was sufficient to provide 9 to 30 years of service, depending on the number of dredging projects required to use the facility. When filled, the site was to be landscaped for use as a public recreation area.

Background

In 1970, the U.S. Congress authorized the deepening of the harbor approach channels from 42 to 50 feet, on the condition that nonfederal harbor agencies provide a suitable placement area for dredged material. In the past, materials dredged from the harbor have been placed at open-water sites. Concerns that this practice would have adverse environmental effects led to an extensive environmental review. Approximately 70 disposal sites were considered, 10 of which were subjected to intensive study. The analysis indicated that the construction of a diked area spanning Hart and Miller islands would provide the most environmentally acceptable disposal site.

At first, the Hart/Miller project was controversial. Proponents argued that it would reduce the open-water placement of contaminated dredged materials and help combat serious erosion of the islands. Opponents asserted that environmental impacts, alternative options, and cumulative impacts had not been assessed adequately. A legal challenge to the plan was filed, but the courts eventually ruled against the plaintiffs. A permit was obtained in 1976, but the legal challenge was not resolved until late 1980. The CDF was completed in 1983 and was first used in 1984.

Remediation Alternatives Considered

In addition to containment at the Hart/Miller site, other strategies that were considered included taking no action, using other diked containment areas (including the use of a larger number of smaller sites), and using dredged material to produce bricks or other ceramics. The range of alternatives was limited when the Maryland General Assembly appropriated $13 million specifically to implement a diked containment strategy. A consultant evaluated 70 sites and concluded that Hart/Miller was the most desirable on the basis of both economic and environmental factors.

Strategy Chosen

The original plan, dredging and disposal at the Hart/Miller CDF, was carried out, and the CDF was operated under rigorous effluent discharge criteria and subject to external environmental monitoring. Subsequently, the dikes were temporarily raised 10 feet, from 18 to 28 feet, for the 50-foot deepening project because substantial capacity was used to contain clean sediments from the approach channel when no other placement sites were available. A continuing inability to secure consensus on placement has resulted in a state decision to raise the dikes permanently in the facility's north cell another 16 feet, which would add 30 million cubic yards of capacity.

Overall Assessment

The CDF was expected to be filled by July 1996. The facility's useful life was shortened because the Maryland Port Administration was forced to use the site for backlogged and new dredging work when permits could not be obtained for open-water sites. The state park encompassing the Hart/Miller site has become a fairly popular recreational destination for boaters. Reactivation of an additional small-scale, shoreside industrial CDF is progressing. The state has not been able to establish other sites because of public and environmental opposition to specific sites, including most beneficial use projects. Consequently, the state has been forced not only to continue to use the CDF for clean sediments but also to increase the CDF capacity despite prior public commitments not to do so.

Lessons Learned

- Stakeholders who do not feel adequately represented in the original decision can delay a project for many years in an effort to ensure that their interests are considered.

- A high-volume CDF has the potential to be a multiuse facility that not only isolates contaminants from surrounding waterways but also, for example, stabilizes adjoining coastal areas and serves as recreational parkland.

- The requirement that all sediments dredged from the harbor, regardless of their quality, be disposed of in the CDF shortened the lifetime of the facility.

- The availability of a large CDF can become a convenient excuse to delay or avoid making politically sensitive, difficult, and controversial decisions to resolve critical dredging problems, shortening CDF capacity for contaminated sediments in the process.

JAMES RIVER, VIRGINIA

Among the contaminated sediment projects reviewed by the committee, the James River case stands out as a clear example of the utility of natural recovery. The James River drains an area of approximately 25,600 square kilometers (km^2) from the hills of West Virginia to the tidal waters of the Chesapeake Bay near Norfolk, Virginia. The river was contaminated as early as 1967 with the pesticide Kepone, which was manufactured until 1975 by a company situated near the upstream limit of the estuarine reach. Commercial fisheries were closed in 1975 because of Kepone contamination.

Background

This estuarine system had been studied extensively prior to the discovery of the Kepone contamination, and these early investigations not only supported the need for subsequent ecological studies but also helped create a strong scientific basis for making sound management decisions. The Kepone distribution pattern coupled with the hydrophobic nature of this pesticide suggested that contaminant transport was dominated by physical processes affecting the distribution of suspended sediments. The dominance of physical processes favored a progressive decrease in surficial Kepone concentrations as a result of dispersion and dilution, as clean sediments introduced from both upstream and downstream sources were mixed with contaminated materials, and as a result of sedimentation and burial within the deeper, low-energy regions of the river channel. These natural processes effectively reduced near-surface concentrations and the associated exposure of biota and isolated contaminants in the interior of the sediment column from overlying waters.

Remediation Alternatives Considered

A variety of options, including dredging, sorption, and stabilization of the surficial sediments using molten sulfur, were considered in a 1978 study. Biological, chemical, physical, and geological aspects of the contamination indicated that remedial actions to remove Kepone would be expensive, time-consuming, and environmentally damaging.

Strategy Chosen

The high cost of active remediation, combined with the observed decrease in Kepone concentrations in surface sediments, favored the selection of natural recovery as the strategy of choice. No direct costs were incurred, although there were economic costs associated with site studies leading to the selection of this option as well as with the closure of the fishery for 13 years while natural recovery was taking place.

Overall Assessment

In this case, natural recovery has proven to be an effective management strategy in both economic and environmental terms. Source control was an important factor in the success of this project (the Kepone manufacturing operation was shut down). Commercial fisheries reopened in 1988 when the contamination decreased, the sediments had been covered sufficiently by uncontaminated materials to diminish the Kepone flux back into the water column, and Kepone concentrations in organisms inhabiting the river were below federal action levels. It is assumed that Kepone concentrations are continuing to decline.

Lessons Learned

- Under the right circumstances, natural recovery can represent the most cost effective, environmentally beneficial, and politically acceptable management scheme.

- Confidence in the decision to allow natural recovery to proceed is contingent on the availability of long-term data concerning the physical, chemical, and biological characteristics of the site.

- Close collaboration between scientists familiar with local site characteristics and the agencies responsible for resource management can contribute significantly to the selection of optimum remediation strategies.

MARATHON BATTERY, NEW YORK

Marathon Battery is an example of the remediation of heavy-metal contamination of a wetland environment within the tidal reach of a river. Marathon Battery is a Superfund site located along the eastern shore of the Hudson River, approximately 80 km north of New York City. The area has a long history of industrial use, first as a foundry producing armaments and then, until 1979, as a plant producing nickel-cadmium batteries. Process waters were discharged into a nearby cove and, through a municipal sewer outfall, into the Hudson, introducing significant quantities of heavy metals, including cadmium, nickel, and lead. Field sampling indicated that the site was contaminated by approximately 50 metric tons of cadmium, the principal contaminant of concern.

Background

After contamination of the area was recognized in the early 1970s, a volume of 90,000 m^3 was dredged and dewatered, and 4,000 m^3 of remaining sediment was placed in a sealed vault. In 1981, the Environmental Protection Agency (EPA) placed the site on the National Priorities List of sites requiring investigation and cleanup under Superfund. After the analysis of field data, the EPA decided to use dredging and chemical fixation to remove 95 percent of the contaminants. The fixed sediments were to be transported for off-site disposal. To facilitate dewatering, treatment, and transport, a multicelled settling basin was designed by the USACE. Plans were disrupted when an historical gun testing platform was discovered in the area that was to be used for equipment staging and waste stabilization. The archeological features of the site were then reviewed and the plans redesigned. In 1992, just prior to the bidding for the remediation contract, the EPA reached a settlement with the parties responsible for the contamination, who agreed to clean up the site to the criteria established by the EPA.

Remediation Alternatives Considered

Only one approach—dredging, treatment, and off-site placement—was considered. The eventual solution differed little from the original plan. The EPA's original plan was to rely on gravitational settling for dewatering the sediments and to use a Portland cement-based process for chemical fixing. The responsible parties proposed using a series of graded screens for dewatering and a proprietary stabilizing agent for chemical fixing. This approach would produce a soil-like material in which the metals would be immobilized.

Strategy Chosen

Remediation of the site, through dredging, treatment, and containment, began in 1993. Stabilized sediments were placed in rail cars for transport to a placement site in Michigan. Actual project costs are not known but are expected to be less than the EPA's original estimate of $48.5 million. An interesting incentive for cost savings was used. Contractors were encouraged to recommend modifications to the original plan; if any accepted recommendations reduced costs, then the savings would be shared by the contractor (40 percent) and the government (60 percent).

Overall Assessment

Remediation began in 1993 and was scheduled for completion by late 1994. But a variety of unexpected problems were encountered. Dredging was slowed by tidal conditions, which limited water depths and occasionally grounded the hydraulic dredge used in the confined inshore areas. Because of relatively high concentrations of coarse sand, gravel, and rock in the deeper areas of the Hudson River, the hydraulic dredge was replaced with a clamshell dredge. Initial dewatering operations were very slow because of persistent clogging of the screens by organic materials, so the process had to be redesigned. Neither the final outcome of the project nor the results of post-project monitoring is available at this time.

Lessons Learned

- Site character and history, including archeology, can have a significant effect on project design and can impede project completion.

- Placement of contaminated sediments at a remote site is both possible and acceptable in some cases.

- Adequate pre-project site assessments are needed to minimize the possibility of costly surprises after the project has begun.

- Incentives for cost savings and other innovations can be included in the contract bidding process.

PORT OF TACOMA, WASHINGTON

The Port of Tacoma project is an example of how stakeholder cooperation and the emergence of a single project advocate can lead to innovative and successful solutions. The Sitcum Waterway, an industrial and commercial shipping channel, is a Superfund site within Commencement Bay in the City of Tacoma. Shorelines of the bay are urbanized, with heavy industry on former tideflats. Less than 2 percent of the original near-shore wetlands remain. Several fish species (including salmon, steelhead trout, sole, and flounder) seek refuge and feed near shore. The release of hazardous substances into the environment has altered the chemistry of the water and sediments. Studies indicated that Sitcum sediments contained elevated concentrations of several metals, organic chemicals, and other contaminants.

Background

In 1983, Commencement Bay Nearshore Tideflats were placed on the priority list of sites requiring investigation and cleanup under the EPA's Superfund authorities. The EPA entered into a cooperative agreement with the state Department of Ecology to conduct an investigation and feasibility study at the site. The EPA's Record of Decision (ROD) detailing the cleanup plan for the bay identified eight problem areas, including Sitcum Waterway. The goal was to achieve sediment quality that would support a healthy marine environment and reduce the risk of exposure from the consumption of contaminated seafood caught in the bay. The ROD also set forth sediment quality objectives. It was agreed that EPA would be the lead federal agency for the remediation of contaminated sediments, whereas the state Department of the Environment would take the lead in controlling the sources of hazardous substances.

The EPA asked the Port of Tacoma to consider including Sitcum sediment remediation as part of longstanding near-shore fill and waterway dredging projects, and the port agreed, assuming the responsibilities of the principal responsible party. The emergence of a single project advocate contributed significantly to dispute resolution and, ultimately, to project implementation. These projects, which were required under the terms of a legal settlement for both navigational and environmental reasons, involved dredging Blair Waterway and filling more than 70 percent of the Milwaukee Waterway to create 24 acres of land for a marine container terminal and wildlife habitat.

Remediation Alternatives Considered

The ROD identified four options, including dredging and CAD (at a cost of $11.5 million), dredging and near-shore containment ($4 million), landfill placement ($20 million), and in-place capping, which was considered unacceptable because it would interfere with vessel operations and long-term maintenance of the channel.

Strategy Chosen

The dredging and near-shore containment option (the least expensive of the four) was selected. The plan satisfied terms of the ROD, which specified that near-shore placement be used only if it could be combined with fill projects that would be permitted anyway for commercial development. The strategy was made possible, in part, because the EPA broadly interpreted Clean Water Act (CWA) restrictions on avoidable discharges of dredged material in waters regulated under the CWA. By satisfying the intent of the CWA and Superfund rather than focusing rigidly on the specific language, the EPA was able to forge a creative solution to otherwise intractable problems.

Overall Assessment

The Blair/Milwaukee/Sitcum project was successful in many ways: It settled Superfund liability for contaminated sediments in near-shore areas, provided sufficient depth for navigation to allow the continuation of port activities, provided cleanup and navigational improvements in adjoining waterways, provided for the environmentally acceptable placement of contaminated sediments, facilitated habitat restoration, and generated filled land in the port for productive (industrial) uses.

Lessons Learned

- The success of a project is directly related to the early identification of all interested parties and concerns connected with previous related projects and on the willingness and ability of all parties to satisfy the concerns of affected stakeholders.

- If the goals of land use, port planning, and resource management can be combined in an environmentally sound and economically attractive plan, then project success is almost assured.

- The availability of on-site space permitting containment or remediation and treatment of contaminated sediments is a significant benefit; it may facilitate project approval by reducing costs and eliminate the need to justify extending the contamination to remote disposal sites.

- Interpretation of requirements based on the intent of the underlying law can eliminate complications and delays and contribute to project success.

WAUKEGAN HARBOR, ILLINOIS

Waukegan Harbor is an example of the ex situ treatment of polychlorinated biphenyls (PCBs). Waukegan Harbor is a Superfund site located on the western shore of Lake Michigan, approximately 37 miles north of Chicago. In the mid-1970s, surveys conducted by the state found significant concentrations of PCBs in the sediment column of the harbor. The PCBs had been used as hydraulic fluids by the Outboard Motor Corporation (OMC) in its manufacturing operations and had leaked into the harbor from 1961 to 1971. Extensive sampling showed that, in the areas within and adjacent to Waukegan Harbor, PCB concentrations ranged from 50 parts per million (ppm) to 520,000 ppm. Navigation dredging was suspended until a remediation project could be completed.

Background

Initially, the cleanup project was simple: Sediments would be removed from the harbor and slip areas and adjoining land sites and placed in a landside containment facility. These plans, developed by the EPA and issued in 1983, were contested by OMC on the basis of costs and the company's lack of involvement. The company's subsequent negotiations with the EPA and the associated litigation and delays led to OMC signing a consent decree and assuming responsibility for completing the cleanup. In the meantime, however, Superfund was reauthorized, and the new legislation forced consideration of advanced treatment to achieve significant contaminant removal rather than simple containment.

Under the new plans, all sediments with PCB concentrations above 50 ppm were dredged and placed in a nearby slip that had been occupied by a recreational marina, which was relocated to another site acquired and donated by OMC. To contain the sediments, a three-foot-thick slurry wall was constructed around the slip. All sediments with PCB concentrations above 500 ppm (a total of about 16,000 tons) were dewatered and removed for treatment using a rotary thermal desorption kiln technology selected by OMC.

Remediation Alternatives Considered

Dredging and containment appears to have been the only approach considered until Superfund legislation dictated the use of advanced treatment technologies. It is not clear whether any technologies other than thermal treatment were considered by OMC.

Strategy Chosen

Cleanup began in 1991. The thermal desorption technology reduced sediment concentrations from 500 ppm to 2 ppm and produced 200 tons of residual PCBs in an oil phase, which required off-site incineration. Cost of the process was approximately $250/yd^3, plus fixed costs of $150/yd^3. Treated sediments were returned to on-site containment cells with bentonite walls. Once dredging was completed, all containment areas were capped with soils and high-density polyethylene. Post-project monitoring indicated some groundwater infiltration in the containment areas, so periodic pumping is required to maintain inward hydraulic gradients.

Overall Assessment

After a convoluted and drawn-out process, the project is now complete. The long delays can be attributed to the litigious nature of the relationship between the EPA and OMC, which stems in part from high costs of thermal treatment and failure to include OMC in the initial planning.

Lessons Learned

- Even seemingly ideal project conditions (e.g., one major contaminant, one responsible party, sediments in accessible sites) do not obviate the need for prudent planning and the development of partnerships among stakeholders.

- Project implementation is facilitated by incentives that encourage voluntary action.

- Unencumbered acquisition and transfer of property rights (for relocation of the marina, in this case) can contribute to project completion; such actions obviously are facilitated by partnerships among, and the early involvement of, all affected parties.

- Monitoring and a long-term commitment to active site management (e.g., through groundwater pumping) are required for permanent remediation.

APPENDIX
D

Using Cost-Benefit Analysis in the Management of Contaminated Sediments[1]

Kenneth E. McConnell

To make decisions concerning contaminated sediments, project managers must weigh the relevant factors and make trade-offs. Just because a treatment technology is available does not necessarily mean it should be used in a given situation. Some technologies are merely prohibitively expensive, especially if large volumes of contaminated sediments are involved, but even after impractical techniques have been eliminated, a single approach must be chosen from a wide range of choices. Making decisions that must take into account multiple viewpoints and factors often necessitates making sophisticated analyses to compare the outcomes of various strategies. This appendix describes cost-benefit analysis, an analytical approach that can assist in decision making and serve as a tool for improving the management of contaminated sediments.

The first section presents an overview of the basis for cost-benefit analysis. The second section, through generic examples, discusses how this tool is applied to the management of contaminated sediments and explores the roles of costs and benefits in decision making. The third section examines the components of costs and benefits and how they are computed. This section concentrates solely on the concepts of costs and benefits, leaving aside complex issues, such as uncertainty, that arise in practical applications. (These issues are addressed in Appendix E). The final section examines practical issues involved in the application of cost-benefit analysis to the management of contaminated sediments.

Many federal agencies have guidelines that explain how costs and benefits are to be computed and used. These guidelines can be applied to making

---

[1] This appendix has been edited for grammar and style; accuracy is the sole responsibility of the author.

decisions about environmental issues but have not been used systematically in the context of contaminated sediments. Cost-benefit analysis can provide valuable perspectives on the best ways to manage contaminated sediments.

OVERVIEW OF COST-BENEFIT ANALYSIS

The allocation of scarce resources is a fundamental issue in public policy analysis. These resources include human labor, the natural assets of the environment and the resource base, and the capital stock created by human effort. Labor, natural resources, and capital are used to create goods and services for consumption in the present and to produce capital stock for the future.

Choosing among alternative plans for the management of contaminated sediments involves trade-offs among the uses of society's scarce resources. The decision to dredge a harbor involves commitments of labor and equipment that cannot be used elsewhere at the same time. Dredging for navigation may entail other uses of the area. A dredged channel might enhance recreational boating, for example, with consequent stresses to the environment, or it might change fishing patterns by replacing shallow water habitat with deeper waters, increasing vessel traffic, and generally encouraging development. Navigation channels could spur reductions in the price of transportation services while also fostering the expansion of the coastal fishing fleet, thereby increasing pressure on fisheries. On the other hand, the decision *not* to dredge a harbor may deprive society of scarce navigation services. Many other trade-offs must be made in choosing tactics and strategies for managing contaminated marine sediments.

When resources are allocated in a given market, the market participants must make these trade-offs; competitive forces typically allocate resources to their most valuable use. But in rivers, channels, and estuaries, where the problems of contaminated sediments arise, market forces allocate resources inefficiently because some of the costs and benefits cannot be captured by individuals or firms. For example, pesticides used for agriculture in drainage basins far from a port can contribute to the contamination of dredged sediments, raising the cost of downstream navigation for those who obtain no benefits from the agricultural practices.

The implementation of management plans for contaminated sediments bestows economic gains on some groups and imparts economic losses to others. Equally important, the absence of management plans and the failure to implement plans also bestows benefits and costs. Taxpayers may benefit by not having to pay for the handling of contaminated sediments, but revenues from a particular fishery may be curtailed by the presence of these sediments. In many cases, the gains of one group greatly exceed the losses of another group. Because the net gains—that is, the social gains—can be quite large, it makes sense to consider in an orderly way whether decisions are socially productive by comparing the gains or benefits from a sediment management plan with the costs.

APPENDIX D

This approach is consistent with guidelines for cost-benefit analysis established by the federal government. One prominent set of guidelines is the *Economic and Environmental Principles and Guidelines for Water and Related Land Resources Implementation Studies* (Water Resources Council, 1983), which reflects more than 50 years of experience in the application of cost-benefit analysis to the allocation of water resources. The *Principles and Guidelines* applies the broad language of cost-benefit analysis to the specific tasks of evaluating water resource projects. The driving force is national economic development.

Cost-benefit analysis is used by various government agencies, such as the Forest Service and the U.S. Environmental Protection Agency. Benefits are prescribed legally in the Oil Pollution Act of 1990 (*United States Code*, Title 33, Section 2701)[2] and the Comprehensive Environmental Response, Cleanup, and Liability Act of 1980 (CERCLA) (42 USC §9601). The growing importance of cost-benefit analysis is described in Smith (1984).

ROLE OF ECONOMICS IN MANAGING CONTAMINATED SEDIMENTS

Generic Example

The role of economics in managing contaminated sediments can be visualized using a simple graph, such as Figure D-1, which shows the basic trade-offs involved in decision making. (This type of graph can be found in leading environmental policy texts, such as Baumol and Oates [1992].) In this stylized view, the horizontal axis indicates the degree to which contaminants in the sediments are reduced. Associated with this variable are certain costs and benefits. The units on the horizontal axis could reflect the percentage of contaminants removed, the cubic yards of sediment removed, or a variety of other measures reflecting a reduction in contamination. For simplicity of discussion, the horizontal axis in Figure D-1 measures the percentage of contaminants removed. At the origin (far left), no contaminants have been removed; this is the level of contaminants prior to removal. At the 100 percent level (far right), all contaminants have been removed.

The vertical axis measures both the costs and benefits of sediment removal. The costs pertaining to contaminant removal include all actions taken to remove and manage contaminants. Each vertical point on the cost schedule (the concave curve) is the least cost of attaining the given percentage reduction in contaminants.

For purposes of illustration, imagine that dredging is the method of removal. The costs of dredging, including the costs of labor and capital to carry out the task as well as the costs of the natural resources or other methods of managing the

---

[2] References to the *United States Code* will be abbreviated using the format: 33 USC §2701.

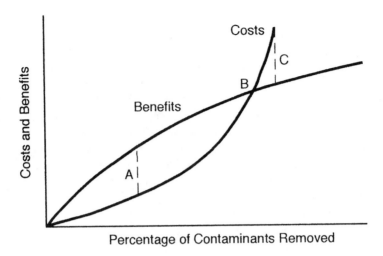

FIGURE D-1 Conceptual illustration of the trade-offs involved in cost-benefit analysis. A = best decision point; B = benefits equal costs; C = worst decision point.

dredged material, rise at an escalating rate as the percentage of contaminants removed increases. Removing the last 1 percent may cost as much as removing the first 99 percent or may not even be feasible. Some cleanup projects, or cleanup to some particular standard, may not be feasible at sites covered by CERCLA, commonly known as Superfund. In these cases, given the degree of cleanup required on the horizontal axis, the costs are so high on the vertical axis that they are beyond consideration.

The benefits include improvements in navigation and in various human services as a result of the reduction in contaminant levels. For example, lowering contamination may mean that the exposure of humans to a potentially damaging substance is reduced or that a closed fishery can be reopened. How these benefits can be measured is explained later in this appendix. For the present, it is sufficient to assume that they can be measured and that all the effects of the contaminants are considered.

As dredged materials are removed, benefits increase and the concentration of contaminants declines. However, the rate of increase in benefits declines (the concave curve). At a certain width and depth of the channel, no additional navigation benefits are provided. At very low levels of contamination, the removal of remaining contaminants may do little if anything to reduce health risks or to improve in ecological functions. This is a classic benefit schedule for the removal of contaminants.

These cost and benefit schedules show how a decision maker can choose the optimal level of contaminant removal. At point A in Figure D-1, the difference between costs and benefits is at a maximum value for the entire graph. This is the best choice from an economic standpoint because the benefits outweigh the costs by the greatest possible amount. For contaminant removal rates to the right of point A, the extra costs exceed the extra benefits, and society is not making the best use of its resources. That is, society is devoting additional resources to removing contaminants but not getting commensurate value. At point B, for example, costs just equal benefits; there would be no net gain from contaminant removal because the gains (measured by the height of the benefit schedule) are just matched by the costs. At point C, the costs of cleanup exceed the benefits, and society has made a poor decision concerning resource allocation. At this point, additional contaminant removal increases costs substantially but provides few added benefits.

When and How Cost-Benefit Analysis Is Applied

The value of comparing benefits and costs stems from the need to make trade-offs. When decisions are made, one type of good or service is substituted for another. When there is a gain in navigation services, there is a loss in scarce factors of production used in dredging and the opportunities for dredging they created. When the scarce factors of production associated with dredging are used to provide an estuary free of contaminated sediments, the value of these scarce resources in other parts of the economy is relinquished. It is a fundamental tenet of economics that factors of production—human services, raw materials, equipment, natural resources—have many uses throughout the economy. When they are used in one part of the economy, opportunities for using them elsewhere are lost. Calculating costs and benefits ensures that when decisions to allocate resources are made, the lost opportunities are counted, and the costs and benefits will not be grossly out of balance. Cost-benefit analysis accounts for the scarcity of the factors of production in terms of their usefulness throughout the economy.

Many objections have been made to weighing costs and benefits in environmental decision making, especially when human health risks are involved. There are two common objections: (1) it is ethically wrong to try to choose sundry economic values over human health; and (2) the benefits encompass many intangible services that are impossible to measure. But strong counterarguments can be made on each score.

Consider the assertion that risks to human health should be minimized, regardless of the cost. This argument ignores the opportunity cost of resources (i.e., the cost of opportunities given up, as discussed later in this appendix). Moreover, even if the goal were to minimize the risk to human life at any cost, that goal could be most effectively achieved by taking some of the resources used to reduce the contaminants from point A to point C in Figure D-1 and investing them in other life-saving programs. With respect to the difficulty of measuring benefits, economists have made considerable strides in measuring the economic value of intangible services, some of which are directly connected with the functioning of marine ecological systems. These advances are described in Freeman (1993) and Mitchell and Carson (1989). Many of the problems inherent in measuring benefits of the marine environment could be resolved through a systematic effort, which might involve gathering simple data on how people spend their time and money—facts that are needed to measure economic values. This type of information can be obtained through straightforward, well-known survey techniques.

The difficulty of computing benefits becomes less discouraging when one considers how cost-benefit analysis is generally applied. In practice, cost-benefit analysis is more likely to be used to compare a small set of projects than a continuum of cleanup possibilities. As long as benefits are calculated consistently, the outcomes will be comparable. For example, it is possible to compare the costs and benefits of several remediation strategies involving different volumes of dredged sediment, although it may not be possible to find consistent measures for comparing the social impacts of dredging to those of, say, incineration. Figure D-2 shows a hypothetical example. Four projects are arrayed in order of the amount of sediment removed. The difference is the net benefit, which, moving from project 1 to project 4, first increases but then becomes negative. (This generic example is comparable to the decision analysis test case described in Appendix E.)

Cost-benefit analysis also can be useful for evaluating targets. Often, remediation projects are designed to remove contaminants to a given level, reduce risks to human health to a given level, or reduce contamination in a frequently monitored species to a certain level. An example of a species-level target would be 4 parts per million (ppm) of polychlorinated biphenyls (PCBs) in flounder. Such targets can be subjected to cost-benefit analysis just as individual practices can. An examination of the costs and benefits associated with particular goals, constraints, or targets can provide two types of insights.

First, by determining whether the benefits of attaining the target exceed the costs, an analysis can indicate whether achieving the target is a good use of scarce resources. Second, the costs and benefits of strategies that achieve different targets can be compared. For example, suppose all the costs and benefits of attaining a 4 ppm "body burden"[3] of PCBs in a critical species (including the economic benefits of the reduced body burden) have been calculated. If the costs exceed the benefits, then it would make sense to examine a less stringent goal, say 8 ppm. Figure D-3 shows two hypothetical projects, one with a target of 8 ppm, the other with a target of 4 ppm. The former is placed to the left because achieving the latter (4 ppm) target requires more resources and involves a greater reduction in contaminant levels. By comparing the net benefits (benefits minus costs) of the two projects, it is evident that, in this hypothetical example, the less stringent target makes more sense economically than the higher target.

---

[3] "Body burden" is defined as the concentration of a contaminant in the tissues of an organism.

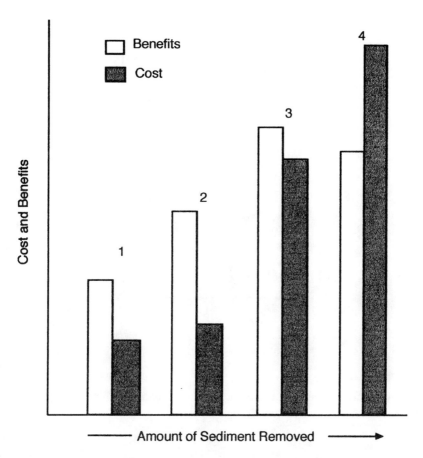

FIGURE D-2 Example of cost-benefit analysis with discrete projects.

CALCULATING COSTS AND BENEFITS

Calculating costs and benefits as part of the contaminated sediments management process does not imply that only costs and benefits matter. Obviously, the social desirability of projects is influenced by many other factors, such as the distribution of costs and benefits among groups with different income levels, legal and regulatory rectitude, and unmeasured but substantiated ecological changes. But it is also important to know how productive, in economic terms, a strategy will be. Calculating the projected economic gains and losses, and distributing this information to the participants in the decision-making process, should help determine the best use of resources. Over time, there will be significant

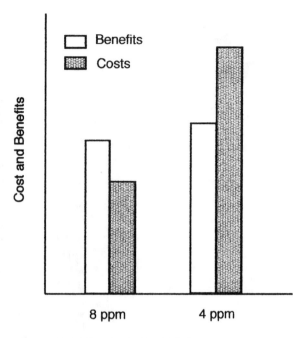

FIGURE D-3 Costs and benefits of reducing body burden.

improvements in resource allocation, and hence increases in real income, when gains exceed costs on numerous projects.

Components of Costs

In practice, the trade-offs among different uses of the marine environment are complex. (These complexities can be handled systematically in decision analysis, as discussed in Appendix E.) Putting aside the complexities for now, there are really only three kinds of costs involved in the management of contaminated sediments:

- dollar costs of remediation and cleanup (dredging, sediment transport, ex situ and in situ treatment, land acquisition, capping, etc.)
- dollar costs of foregone port services as a consequence of capacity constraints and channel restrictions
- environmental costs (in dollars) of contaminants in the sediments (including damages to natural resources from foregone use of the natural environment as well as the costs to human health from exposure to contaminants)

The costs of remediation and cleanup are the most obvious because they entail out-of-pocket expenses and, therefore, are likely to be the first costs considered. These costs pertain to any action taken to reduce the harm from and exposure to contaminated sediments.

The second type of cost is associated with the impact of contaminants on ports. These costs reflect the value of opportunities foregone as a consequence of the contaminated sediments. Transportation and port services can be curtailed in several ways. Failure to maintain channels by dredging restricts shipping. Ships may be able to enter only at high tide, or they may have to lighten their loads by transferring cargo to other ships outside of port. Restrictions, delays, and extra handling add costs to the movement of cargo through the port. Ultimately, these costs are absorbed by the general public in the form of higher prices for transported goods. Port services are also impaired by barriers to the expansion of port capacity through the building of extra piers, slips, or other elements of port infrastructure. These costs are important but are probably the least understood, and the most difficult to quantify.

The broad area of environmental costs encompasses several components. The best-quantified aspect is the damage to natural resources, which is roughly equivalent to the economic losses incurred as a consequence of injury to natural resources. In the case of contaminated marine sediments, damages are caused by the presence or resuspension of contaminants, which may alter the benthic ecology, change patterns of food availability for particular species of fish, or injure commercially or recreationally valuable species throughout the food chain. A once-viable fishery could be eliminated, or restrictions might be imposed on the catch for health reasons. In addition to damages to natural resources, there are ecological effects and direct costs from the health effects of contaminated sediments. In some instances, exposure to contaminants increases the risk of morbidity or mortality in humans. In a statistical sense, the excess risks of morbidity and mortality imposed by exposure to hazardous or contaminated sediments are part of the costs. The category of ecological effects is a catchall and reflects the fact that not all costs can be measured. Ecological functions of marine resources can be degraded in many ways; for example, the environment might lose its capacity to support rare or endangered species protected for the sake of ecological diversity rather than economic value. Only some types of degradation can be measured, and only a subset of these is amenable to economic valuation.

Defining Costs and Benefits

The components and logic outlined above apply not only to costs but also to benefits because the benefits of an action are simply the costs of not taking the action. One of the advantages of considering costs and benefits explicitly in public decisions is the conversion of different services into the common unit of money. But a common unit does not, by itself, ensure comparability. Costs and benefits of different management strategies and services must also be based on the same concepts. Whether the focus is on the treatment or removal of sediments, natural resource damages, or other aspects of costs, a common conceptual framework guides calculations. Only when the same ideas motivate the calculations of the costs is it reasonable to discuss trade-offs among actions. A consistent conceptual framework is outlined here.

Economic benefits of a project, service, or access to a resource can be defined as the public's maximum willingness to pay for the project, service, or access to a resource rather than do without it. The public's willingness to pay is simply the sum of the willingness of private individuals to pay. Costs need to reflect opportunities given up. In other words, the cost of a certain action is the dollar value of the best alternative course of action that could have been pursued instead. This is the fundamental definition of cost as opportunity cost. For example, the cost of using an acre of land for the disposal of dredged material is the dollar value of the next-best alternative, which in most cases is the price of that acre of land. The cost of hiring an engineer for two months on a project is the opportunity cost of employment elsewhere, which would be the engineer's wages in most cases. The cost of preparing a site to receive contaminated sediments is the cost of labor and equipment that could have been used elsewhere. The notion of opportunity cost can be used whenever there is any doubt about exactly what the costs are.

The use of opportunity cost as the fundamental basis of costs connects costs and benefits. Benefits are simply the reverse of costs: The benefits of an action are the costs of not taking the action. For example, suppose that the lost value of recreational use of a beach is one of the costs of not removing contaminated sediments. Then one of the benefits of removing the sediments is the incremental value of recreational use due to the removal. The arithmetic of cost-benefit analysis is illustrated in Box D-1.

For the purposes of calculation, it is also essential to define which costs and benefits matter. For public decisions, social benefits and social costs are the key measures. These are any costs incurred (or benefits received) by anyone affected by actions (or the lack of action) concerning contaminated marine sediments. Social costs include all the usual private costs as well as costs not typically thought of as private. For example, a port pays the private costs of operating a dredge but does not cover the health and material expenses inflicted on others or on the ecosystem at the sediment placement site (although in some cases these costs are paid in mitigation elsewhere). The social cost is the sum of the private costs and the external costs (adding contaminants to the placement site). Alternatively, when a firm uses a toxic substance, it pays the material costs through the purchase but may not pay the costs imposed on others if the substance escapes into the marine environment. The social cost of the use of the toxic substance is the sum of both costs. The social benefits are the private benefits that accrue to ports and their users and are transmitted through the pricing system.

Economists always take a human perspective in discussing costs and benefits. Only the costs that humans incur or the benefits that accrue to humans, currently or in the future, count in weighing costs and benefits. The human perspective on costs and benefits reflects the role of humans as decision makers. Humans cannot avoid the consequences of their decisions and so take them into account as explicitly as possible. This anthropocentric viewpoint makes some observers uncomfortable. What about future generations or nonhuman aspects of the natural environment, both now and in the future?

Critics often believe, mistakenly, that these issues are ignored, simply because outcomes are valued monetarily. The mistake lies in equating costs and benefits with market outcomes. Just because economics measures the value of alternatives based on monetary returns to humans does not mean that future generations do not count or that nonhuman elements are not important. Those elements count to the extent that humans want them to count. Society can register concern for future generations by preserving resources or providing productive capital for the future. Humans demonstrate concern for the natural environment by protecting it, even when this effort requires giving up control over resources that could be consumed or used for production. In decisions concerning environmental issues, this concept is often introduced when adverse impacts are found to be unacceptable. Some adverse impacts are acceptable, and the features that make them acceptable can be incorporated into the cost-benefit analysis. When cost-benefit analysis is used properly, all aspects of the natural environment are considered, but from the human perspective.

Costs and benefits incurred at different times have different values. A treatment process that costs $1 million has a different value today than one that will cost $1 million in two years. The difference is due to discounting: the notion that $1 one year from now is worth (or costs) $1/(1+$r$) today, where r is the appropriate interest rate (or the discount rate). The present value of the cost of $1 incurred 20 years from now is $1/(1+r)20$. In the calculation of natural resource damages, which may be incurred for decades, the role of time is critical. Discounting and the selection of the discount rate have a substantial bearing on the magnitude of costs and benefits.

The discount rate measures society's consensus concerning the value of postponing the use of a resource for a period of time. Thus, the discount rate is a critical consideration when decision making is delayed or has long-term effects. When decisions have long-term impacts, as is the case when contaminated sediments are deliberately left in place, the discount rate has a strong influence on costs and benefits; higher rates basically reduce the influence of future generations.

In considering the value to society of public projects, a common mistake is to confuse economic impacts with costs and benefits. Economic impacts measure the dollar value of market transactions, such as beachfront rentals and hotel and restaurant revenues. Benefits minus costs (or economic welfare) reflect society's net change in well-being and measure the value to society of what is obtained,

> **BOX D-1**
> **Simplified Examples of Cost-Benefit Calculations**
>
> Economic gains and losses are just other words for costs and benefits. To calculate net benefits, one subtracts costs from benefits. In the context of contaminated marine sediments, the costs and benefits are measured by how activities, such as port use and recreation, are influenced by the presence of the sediments and by the health and ecological effects of the contaminants. The use and accounting of costs and benefits are illustrated in the following examples.
>
> Consider two strategies for managing contaminated sediments in a waterway where dredging is necessary for navigation. One alternative foregoes dredging to cap the sediments at a cost of $3 million. Because there is no dredging, no navigation services are available. The second alternative dredges the sediments and removes them for ex situ treatment at a cost of $10 million. But, as a consequence of the dredging, $12 million in navigation services becomes available.
>
> The first alternative has a net gain of – $3 million (i.e., the benefits, which are zero, minus the $3 million cost of capping). The second alternative has a net gain of $2 million ($12 million gain in navigation services minus the $10 million cost of dredging). If the plans were equal in all other respects, then the best alternative would be dredging with ex situ treatment.
>
> In another example, suppose that dredging allows navigation but resuspends contaminants, and as a consequence eliminates recreational fishing. Without the dredging, the value of recreational fishing is $5 million. The dredging costs $1 million and permits an incremental gain in navigation services worth $2 million. Suppose the first alternative is natural restoration. The benefit of this plan is $5 million, the value of the recreational fishing. The dredging alternative is worth $1 million ($2 million gain in navigation services less the $1 million cost of dredging). In this example, natural restoration would be the preferred strategy from the perspective of gains and losses.
>
> The second example can be viewed in a different way without changing the substance. When the strategy of natural restoration is adopted, navigation services are foregone and hence may be considered a cost of natural restoration. But navigation services cannot then also be a benefit of dredging; that would be double counting. In the same way, the foregone value of recreational fishing could be considered a cost of dredging but then could not be counted as a benefit of natural restoration. The

> important point is, whether such values are counted as costs of one alternative or benefits of another, there is no impact on the result of the analysis. This is evident from a comparison of the two ways of viewing the second example:
>
> **Navigation counted as a benefit of dredging**
>
> A. The net benefits of dredging are $2 million benefits to navigation −$1 million cost of dredging
> B. The net benefits of natural restoration are $5 million benefits from fishing −$0 cost
>
> $B - A = 3 - (-1) = 4$. **Choose B. The result is the same as above.**
>
> **Navigation counted as a cost of natural restoration**
>
> A. The net benefits of dredging are −$1 million cost of dredging
> B. The net benefits of natural restoration are $5 million benefits from fishing −$2 million value of lost navigation
>
> $B - A = 3 - (-1) = 4$. **Choose B. The result is the same as above.**

over and above the value of what must be given up. Economic impacts do not measure the net value of projects, and, in any case, impacts are often costs, not benefits.

There is a close connection between economic impacts and so-called secondary benefits. Secondary benefits encompass the additional spending generated from initial expenditures. Spending does not stop with the paycheck for an engineer hired to help dredge a port, for example. The engineer spends income on food, housing, and other services. This spending, however, constitutes economic impact only, and not additional economic value, because the spending could have been generated just as well by dredging at another port. Economic impact is not the focus of cost-benefit analysis, which attempts to measure the net increase in economic value resulting from increased efficiency in the allocation of resources.

Social costs and benefits can be distinguished from secondary effects and impacts using a simple test. If a household, firm, or economic agent is influenced directly, through a physical process or fear of such a process rather than through the market, then it is considered a gainer or loser and belongs in the social accounting. When a household or firm is influenced through the market as a result of someone else's direct response, then the reactions are secondary, or economic, impacts. For example, suppose a notice of contaminated sediments reduces recreational fishing in an area. A true social cost is the loss to recreational fishermen. This cost can be measured using various techniques outlined by Freeman (1993) that involve finding evidence for the reduction in anglers' monetary valuation of recreational opportunities. An economic impact is the reduced spending at the bait and tackle shop in the vicinity of the contaminated sediments. The bait shop operator may suffer, but someone else benefits as a consequence of the redirected spending of the recreational angler. The redirected spending represents a transfer, not a net increase or decrease in social benefits. The only certain, clear cost is absorbed by the recreational anglers.

Costs of Remediation and Cleanup

In the context of the management of contaminated sediments, most discussions of costs begin with disposal costs. Although there is considerable uncertainty about the magnitude of these costs, few ambiguous or controversial issues are involved in calculating them. These costs can be calculated based on market prices, and they involve tangible goods or services, such as wages and salaries, purchases of raw materials, rentals of equipment, and purchases of land. The calculation of these costs is an application of cost accounting. The uncertainty stems from lack of experience with different scales of operations.

Measuring the Economic Cost of Constrained Port Capacity

Many of the current cases of contaminated sediments involve ports. In a typical situation, port managers want to dredge to maintain or increase channel capacity. The process of dredging and managing sediments creates a threat to the environment and public well-being, a threat that can be considered an economic cost. (Economic costs are examined in the next section.) But without dredging, port capacity may be restricted to the point that higher costs are imposed on users of the port. With dredging, costs are lowered, thereby providing economic gains to port users and, hence, to society at large.

The U.S. Army Corps of Engineers (USACE) has dealt with many of the issues that typically arise in addressing the opportunity costs of not dredging. The USACE generally is organized to consider the costs and benefits of new-construction dredging projects, such as dredging a new channel. In its project analysis, the USACE would consider the benefits of the project as navigation benefits and the costs as the direct costs of dredging and similar activities.

To compute navigation benefits, the USACE uses the *Principles and Guidelines* described earlier. There are two types of navigation benefits: benefits related to inland navigation and benefits for deep-draft navigation. Whereas the specific issues that arise for inland and deep-draft navigation may differ, the principles of benefit calculation are the same. According to the *Principles and*

Guidelines, "The basic economic benefits from navigation management and development plans are the reduction in the value of resources required to transport commodities and the increase in the value of output for goods and services" (Water Resources Council, 1983). This definition is consistent with standard applications of cost-benefit analysis under most circumstances. In most applications, navigation benefits are simply the cost savings from reductions in restrictions.

There are two types of cases in which USACE calculations of navigation benefits are incomplete. The first category includes cases in which additional traffic is induced at one port at the expense of traffic at another port; this situation is covered in the *Principles and Guidelines* but is not addressed in practice. The second situation arises when national policy affects prices for navigational services, in which case the effects of price changes, particularly on other port facilities, must be considered. Current USACE procedures for cost-benefit analysis do not account for possible changes in transportation prices as a consequence of investment projects. It is beyond the task of this committee to consider the economic effects of such large projects, but it is clearly an important issue. The problems of cost-benefit analysis in the case of price changes are discussed in Just et al. (1982).

Understanding and Measuring Environmental Costs

The growth of the environmental movement, as well as increased understanding of the ecological and environmental effects of the dredging and storing of sediments, has changed the cost-benefit calculation by introducing a third component, known broadly as environmental costs. These costs reflect injury or the threat of injury to a resource or to users of a resource. Recreational users may be prevented from using a certain site, for example, or households may incur health risks through the consumption of contaminated fish products, or the ecological productivity of a wetland may be impaired by contaminated sediments. These are all examples of environmental costs of contaminated sediments.

Considerable progress has been made in developing methods for calculating natural resource damages. There is a wealth of work on natural resource damages in the context of Superfund, as amended, and the Oil Pollution Act of 1990, but the issue in the present context is more about the lack of understanding of the kinds of services that are lost and how they can be valued monetarily than it is about legalities. Kopp and Smith (1993) and Ward and Duffield (1992) provide valuable background on this subject.

Natural resources generate two kinds of economic value: use value and nonuse (or existence) value. Use value, a widely accepted and frequently used measure, is simply the economic value provided by the opportunity to use resources for recreation, commercial fishing, and other direct uses. Nonuse value is the economic value of goods and services that a person who has no plans for using the resource, currently or in the future, will give up in order to preserve the resource in its current state. The idea of nonuse value, and especially the methods used to measure it, are subject to controversy (addressed in *Federal Register*, vol. 58, no. 10, January 15, 1993, pp. 4601–4614).

A recent study of the economic losses attributed to contaminated sediments focused on the presence of PCBs and dichloro-diphenyl-trichloroethanes (DDTs) in the coastal waters off Los Angeles. These chemicals had various adverse ecological effects. Researchers estimated the present value of the economic losses per English-speaking household in California to be about $55 (Carson et al., 1994). When this amount is multiplied by the more than 10 million English-speaking households in California, the total losses come to approximately $575 million. Because most of these households never use the marine waters, it is reasonable to conclude that these losses would be passive use or nonuse values.

Consider the case of PCBs in New Bedford Harbor. The chemicals, which originated from various manufacturing plants, were deposited into the Acushnet River and the harbor over a period of years. As evidence of the risks from PCBs became known, the Superfund provisions for the recovery of natural resource damages were implemented. Federal and state governments sued the principal responsible parties for damages resulting from the PCBs. Two kinds of uses were impaired: recreational use and the general enjoyment of housing services for waterfront or near-waterfront housing. The housing services are discussed in Mendelsohn and Huguenin (1992).

The intuitive meaning of the damage calculation emerges in the broad context of resource allocation. The figure of $3 million in damages over a 20-year period implies that households would be willing to give up control over valued resources (as measured by their own incomes) to prevent or eliminate PCB contamination at several beaches. Thus, if it were discovered that remediation costs were well in excess of $3 million, then such expenditures would have to be justified on grounds other than the efficient use of resources for households. The lost value of housing services was estimated to be more than $30 million (Mendelsohn and Huguenin, 1992).

There are two additional components of environmental costs: health costs and ecological costs. Health costs are the costs of increased morbidity and mortality resulting from exposure to contaminated sediments. The calculation of these costs is explained in considerable detail in Freeman (1993). The increased morbidity imposes costs through lost work, pain of illness, and medical costs. The costs of mortality are measured by studies demonstrating what society would pay in social terms to prevent an increase in mortality in a statistical sense. This approach involves assigning a value to a statistical life. For example, a substance that imposes a human risk of mortality of 1 in 10,000 would induce 5 statistical mortalities if 50,000 people were exposed. Given an estimate of society's willingness to pay to prevent the loss of statistical life, the benefits of preventing the exposure to the substance could be calculated. Evidence that contaminated sediments induce morbidity or mortality is rare, although not unheard of, so these measures typically are not needed to assess management plans. A discussion of the economic costs of increased morbidity and mortality can be found in Cropper and Freeman (1991).

The last component, ecological costs, includes the economic costs of damage to the ecological functioning of a natural resource. These costs can sometimes be measured, for example when wetlands provide habitat for a species that is harvested commercially. In most cases, however, the term "ecological cost" is an admission that some aspects of the environmental effects of policies cannot be measured and must be assessed qualitatively. Examples of cases in which the value of some service can be measured include the *Exxon Valdez* spill and the PCB-DDT contamination of the Los Angeles Bight.

USING COST-BENEFIT ANALYSIS IN DECISIONS ABOUT CONTAMINATED SEDIMENTS

The discussion so far has dealt with the role of cost-benefit analysis in resource allocation and the nature of costs and benefits in the context of contaminated sediments. An obvious practical question at this point is how cost-benefit analysis can be used in the decision-making process. It is useful to think of applying these ideas at two levels. On the broad level, decisions about contaminated sediments are part of a larger set of decisions about scarce resources, and in that context it is reasonable to ask whether the benefits to society that result from the decisions warrant the costs to society.

On the more concrete level, the role of cost-benefit analysis depends on the nature of the decision and on the legislative and regulatory setting. Some type of cost-benefit analysis is warranted in any decision concerning contaminated sediments. The point is to introduce into the debate a particular way of thinking on the part of project proponents, opponents, mediators, and other interested parties. Costs and benefits must be considered and weighed seriously by decision makers. When the costs exceed the benefits, there must be cogent reasons to proceed with the project, which is not, in broad economic terms, in the public interest.

Cost-benefit analysis also has a role in large maintenance dredging projects and new-work projects. The USACE guidelines require cost-benefit analysis for new projects. But many controversial projects involve maintenance dredging. Dredged sediments now have to meet biological criteria, which depend on density but not on the dispersion of toxic materials. For very large maintenance dredging projects, social decision making can be improved if costs and benefits are considered as well as whether the dredged sediments meet the specified criteria.

SUMMARY

This appendix has outlined the role of economic principles in the choices that confront decision makers and society in determining the efficacy of dredging, the disposition of dredged material, the nature and extent of cleanup, and the general nature of social trade-offs in the managing of contaminated sediments. The basic principle is that activities should be undertaken if the social gain, correctly measured (i.e., in an acceptable manner), exceeds the social cost, correctly measured. These measurements become particularly important when the stakes are large.

There are, however, some abiding themes that suggest likely qualitative relationships, even for small projects in which the measurement of costs and benefits does not seem practical. First, the cleanup of contaminated sediments tends to become increasingly costly as the concentration of contaminants declines. Furthermore, the social gains from cleanup tend to increase ever more slowly as the concentration of contaminants declines. However, there are situations in which certain initial measures taken to reduce contamination could be relatively inexpensive, whereas the corresponding social returns could be quite high or vice versa. Decision makers need to take due account of such considerations in weighing costs and benefits.

REFERENCES

Baumol, W.J., and W. Oates. 1992. The Theory of Environmental Policy, 2nd. ed. Englewood Cliffs, New Jersey: Prentice-Hall.

Carson, R.T., W.M. Hanemann, R.J. Kopp, J.A. Krosnick, R.C. Mitchell, S. Presser, P.A. Ruud, and V.K. Smith. 1994. Prospective Interim Lost Use Value Due to DDT and PCB Contamination in the Southern California Bight. La Jolla, California: Natural Resource Damage Assessment, Inc.

Cropper, M.L., and A.M. Freeman, III. 1991. Environmental health effects. Pp. 165–211 in Measuring the Demand for Environmental Quality. J.B. Braden and C. Kolstad, eds. New York: Elsevier Science Publishing Company.

Freeman, A.M., III. 1993. The Measurement of Environmental and Resource Values. Washington, D.C.: Resources for the Future.

Just, R.E., D.L. Hueth, and A. Schmitz. 1982. Applied Welfare Economics and Public Policy. Englewood Cliffs, New Jersey: Prentice-Hall.

Kopp, R.J., and V.K. Smith, eds. 1993. Valuing Natural Assets. Washington, D.C.: Resources for the Future.

Mendelsohn, R.O., and M. Huguenin. 1992. Measuring hazardous waste damages with panel models. Journal of Environmental Economics and Management 22(3):259–271.

Mitchell, R., and R. Carson. 1989. Using Surveys to Value Public Goods. Washington, D.C.: Resources for the Future.

Smith, V.K., ed. 1984. Environmental Policy under Reagan's Executive Order: The Role of Benefit-Cost Analysis. Chapel Hill: University of North Carolina Press.

Ward, K.M., and J.W. Duffield, eds. 1992. Natural Resource Damages: Law and Economics. New York: John Wiley & Sons.

Water Resources Council. 1983. Economic and Environmental Principles and Guidelines for Water and Related Land Resources Implementation Studies. Washington, D.C.: U.S. Government Printing Office.

APPENDIX
E

Using Decision Analysis in the Management of Contaminated Sediments[1]

John Toll,[2] Spyros Pavlou, Dwayne Lee, Larry Zaragosa,[3] and Peter Shelley

There are many ways to make decisions about how to manage contaminated sediments. Often the process is dictated by legal or political realities. In other cases, the process depends on the complexity of the decision. Simple decisions can be made almost instantaneously, whereas slightly more complicated problems may require a calculator or a detailed examination. Highly complex decisions, involving large amounts of data and considerable disagreement and uncertainty, may require more structured methods to account for all important factors. Computational or computer-based decision support is useful when the issue at hand is of substantial importance and uncertainty, when the situation is politically or emotionally volatile, and when the outcome must be acceptable to all parties. The management of contaminated sediments often falls into this category.

Cost-benefit analysis (addressed in Appendix D) is one tool for making complicated decisions, but it may not be sufficient in some cases. What is needed is a reliable tool for balancing the consideration of a variety of significant factors when the stakes are high and the issues are complex. Decision makers need to know how to use and communicate information about risks, costs, and benefits—information that may be controversial and difficult to evaluate, compare, or reconcile. This issue, identified earlier by the National Research Council (1989),

---

[1] This appendix has been edited for grammar and style; accuracy is the sole responsibility of the authors.

[2] Parametrix, Inc., Kirkland, Washington.

[3] Environmental Protection Agency, Office of Solid Waste and Emergency Response, Washington, D.C.

was assigned specifically as a task to be addressed in the present report. The need is becoming increasingly urgent as the number of remediation proposals grows and as costs and controversies multiply. To add to the burden, some of the committee's proposals for improving outcomes may introduce additional considerations that may initially complicate decision making.

One tool that may help resolve problems with many variables is decision analysis, a computational technique for predicting the outcomes of selected management approaches. Decision analysis provides a way to use both factual and subjective information to evaluate the relative merits of alternative courses of action. Decision analysis does not provide absolute solutions, but it can offer valuable insights. It can integrate the results of key management tasks (site characterization, risk assessment, technical feasibility studies, and economic assessment) into explicit models of the problem as it appears from the perspectives of different stakeholders. The modeling approach allows stakeholders to explore disagreements about subjective elements of the problem, thereby expediting problem solving. The process also formally accounts for uncertainties.

Although decision analysis involves complex computations, the general process can be described in simple terms. The process begins with the gathering of information about the problem and the selection of alternatives to be evaluated by the mathematical model. The model evaluates and rates all possible outcomes to each alternative. The model then identifies the alternative with the highest expected net benefit—that is, the strategy that offers the best odds for successful risk management although it cannot guarantee the best outcome.

Although decision analysis is not a new technique, it is only beginning to be used in managing contaminated sediments. Apparently, the first use of decision analysis in the management of contaminated sediment was in 1996 by Parametrix, Incorporated, in the case of the Asarco smelter site on Commencement Bay, Washington. Such applications may be particularly timely now because recent advances in computer hardware and software have made user-friendly, interactive analyses possible. Further impetus may be provided by the U.S. Congress, which is considering requiring formal risk-based assessments and cost-benefit analyses of proposed federal environmental regulations with projected annual costs of $25 million or more.

This appendix examines how decision analytic techniques can be used to weigh economic, human health, and environmental risks and benefits in a balanced way. The first section provides a general introduction, including background on the unique benefits of decision analysis, the technical basis and merits of balancing risks, and the state of practice. The second part summarizes the practical benefits of decision analysis as demonstrated in a hypothetical test case developed by the committee using actual field data. The test case, which involved choosing the best of three dredging and placement strategies for a hot spot contamination site, is described in detail in the third section.

PART I: BACKGROUND

It is important to distinguish decision analysis from the many other decision-making approaches that have emerged in recent years to improve the process of dispute or conflict resolution. The simplest method is to bring stakeholders together for a frank and constructive discussion. Other approaches include mediation, negotiated rule making, and collaborative problem solving. These approaches may be easier to explain to stakeholders and may, therefore, be approached with less skepticism than decision analysis, which is technical in design and involves complex computations. In some cases, they will be complementary to decision analysis. Each approach has a place in the arsenal of techniques that can be used to improve the prospects of making a politically acceptable and implementable decision. A detailed analysis of all the techniques is beyond the scope of this appendix, but the key aspects are summarized here.

Fostering a consensus on a management process among all interested or potential stakeholders is a separate discipline from decision analysis, and, whether carried out in the context of mediation, arbitration, or collaborative problem solving, it is more than simply going through the mechanics of communicating with all parties. The case studies reviewed by the committee (see Appendix C) underline the paramount importance of positive working relationships in fostering progress toward accommodating or resolving conflicts.

The literature on conflict resolution, in addition, stresses that the way threshold questions are handled is as central to success as the substantive outputs from the process itself. Threshold questions include who should be at the table, who should represent whom, how the interests of important stakeholders who fail to come forward will be determined, how to develop a common and constructive definition of the problem(s), and how to select a mutually acceptable decision-making process. There is a significant body of literature on dispute or conflict resolution. Carpenter and Kennedy (1988) provide lay readers with an extended discussion of the mechanics of a powerful dispute resolution program, many examples of public dispute resolution, and a detailed bibliography. Another resource of this type is Singer (1990).

Decision analysis, as defined and developed in the present report, is not intended to be and is not directly applicable as a dispute resolution technique. Increasingly, contaminated sediments are managed in complex, ever-changing social and political settings marked by the emergence of nontraditional stakeholders in addition to project proponents and regulatory agencies. Conflict is inevitable in this context regardless of the quality of the decision, and serious disagreements need to be addressed directly through appropriate conflict or dispute resolution. Federal agencies are authorized and encouraged to engage in alternative dispute resolution techniques by the Administrative Dispute Resolution Act of 1990 (P.L. 101-552). The U.S. Army Corps of Engineers (USACE) has developed guidelines for using these techniques to resolve contract disputes but has not formalized their use in situations involving contaminated sediments. The Environmental Protection Agency (EPA) frequently uses formal dispute resolution techniques, both for developing regulations and for resolving disputes in specific Superfund projects.

Decision analysis is unique among the available techniques in terms of how it can improve the understanding and quality of decisions and choices. It is explicit and rigorous, and the analytical pathways are reproducible. Decision analysis has the particular virtue of integrating data and expertise from divergent sources into a single analysis, accommodating more variables and offering different perspectives than techniques such as cost-benefit analysis that evaluate single outcomes. Because of its strengths in handling multivariate problems and its capability to model various outcomes (so that the consequences of differing values or assumptions can be tested), decision analysis can be a powerful tool for conflict or dispute resolution in both public policy and project-specific settings. However, because the procedure is elaborate, stakeholders may need some time, as well as demonstrations, to gain confidence in the approach.

Risk Balancing

Uncertainties and disagreements concerning risks, costs, and benefits are impediments to effective decision making. Uncertainty can foster risk aversion and polarize already-divergent opinions. If the uncertainties are not explicit and available to the decision maker for analysis, then balanced decisions can be very difficult to make—and, perhaps, even more difficult to explain to the satisfaction of stakeholders. Decision analysis is a systematic approach that rigorously accounts for uncertainties and disagreements. If done well, decision analysis instills discipline in the overall problem-solving process, forces stakeholders to be explicit about their value assumptions, and provides disciplined consideration and interpretation of information about risks, costs, and benefits.

The merits of reliable risk balancing through decision analysis are fourfold. First, decision models can systematically work through calculations (arithmetic or logical) that are far too complex to perform manually in a timely and orderly way. Second, decision models can be used as records of how problems are formulated. The record can be communicated, examined, and critiqued so that the final model for a specific problem contains the collective insight of a variety of individuals, each of whom may have specialized knowledge and a different perspective. Third, the collective modeling effort and availability of a record may increase confidence and trust in decisions among those whose knowledge or concerns should be addressed in the decision-making process. The model can clarify the implications of uncertainties about factual information and disagreements about subjective aspects of the problem. The ultimate result can be a cooperative problem-solving environment, consensus building, and the expeditious

APPENDIX E 261

implementation of solutions. Finally, by accelerating the evolution of understanding, decision modeling can lead to faster and better solutions than would otherwise be possible. Formulations of the model can be run early in the decision-making process to test alternative assumptions and to evaluate the importance of various uncertainties.

State of Practice

Decision analysis was developed in the late 1960s and early 1970s (Raiffa, 1968; Keeney and Raiffa, 1976) and has been used to help solve policy, management, engineering, and medical diagnostic problems since the 1970s and is increasingly becoming a part of academic curricula in these fields (e.g., Howard and Matheson, 1984; Watson and Buede, 1987; Baird, 1989; Clemen, 1991). Recent advances in personal computers have spawned a new generation of powerful software packages that provide user-friendly, diagrammatic interfaces for building models of decision problems, graphics tools for analyzing modeling results, and rapid processing of computationally intensive problems. These developments have opened up new possibilities for applying decision analysis to a broad array of difficult problems, including environmental management problems. For example, decision modeling software enabled the committee to construct and display models as diagrams and spreadsheets, rather than as lines of computer code, and to display model output as useful graphs and diagrams (some of which are included in this appendix). These figures made it possible to perform valuable exploratory analyses of the problem in ways that would have been impractical without the software.

Few environmental management applications of decision analysis are well documented in the scientific or professional literature, although a number have been identified by the committee, particularly in the public sector. Decision analytic tools have been used successfully by a variety of federal agencies. An example is the development of National Ambient Air Quality Standards (NAAQS) by the EPA. Decision analysis was considered useful in that process because of the significant economic impacts of alternative standards and the complexity of evaluating the available scientific literature on the health effects of NAAQS (EPA, 1996). Development of the NAAQS for ozone, a primary motivation for pollution-control programs, such as vehicle inspection, was supported by the use of decision analytic methods for arraying the health impacts of alternative specifications.

Decision analytic methods also have proven useful for evaluating options for dumping of low-level radioactive wastes in the ocean. Information comparing different management choices has been valuable to the United States at international meetings such as the London Convention of 1972.

Some of the most significant efforts to use decision analytic techniques have been related to the environmental restoration of federal facilities. The costs of cleaning up these sites are significant—hundreds of billions of dollars under some scenarios. Although decision analysis has proven useful for organizing information (U.S. Department of Energy, 1986), it has not been as successful for developing NAAQS, for example, for at least two reasons. First, the use of decision analytic techniques was an emerging concept at that time, and there was little precedent for its application to a wide range of problems. Second, it is not clear that the methodology (as applied at that time) adequately addressed the differences in values and uses of information desired by decision makers.

The U.S. Department of Energy (DOE) made another attempt to use decision analytic techniques for environmental waste management with its Response Allocation Resource System model. This model was designed to support setting priorities among projects to be funded by DOE's waste management division. The model was expected to provide insight concerning public risk, worker risks, environmental risks, and the costs of compliance with environmental statutes and to help set reasonable schedules and budgets. Several public meetings were held and scientific reviews conducted, but the project is, at this writing, still on hold.

Elsewhere in the public sector, Los Alamos and Sandia National Laboratories are using decision analysis to evaluate the costs, risks, and benefits of all their environmental, safety, and health action plans (D. Brooks, Arizona State University, personal communication to Marine Board staff, January 18, 1995). Decision analysis also has been used to analyze the societal risks, costs, and benefits of regional strategies to reduce ozone in Southern California. In addition, the Northwest Power Planning Council uses decision analysis in its activities, which require making multibillion dollar decisions based on forecasts of demand, costs, technologies, the political feasibility of particular alternatives, and environmental predictions that extend decades into the future (Northwest Power Planning Council, 1991).

There are few published examples of decision analysis in the private sector. Chevron Corporation uses decision analysis to help integrate the quantitative analysis of environmental risks into its management decisions, and DuPont Environmental Treatment has used a decision analytical approach to find cost-effective, environmentally sound wastewater treatment methods (Horton, 1993). Details on projects like these are seldom available to the public.

As practical applications of decision analysis become more numerous in environmental decision making, the general concept is receiving attention in the literature as well, although the terminology varies. For example, the concept of decision analysis provides the underpinning of recent EPA guidance on data collection in support of environmental decision making (EPA, 1994). However, the term "decision analysis" is not used. The term is used in another conceptual guidance document, a recent Chemical Manufacturers Association report on the use of decision analysis for characterizing quantitative cancer risks (Silkien, 1990).

PART II: ASSESSMENT OF DECISION ANALYSIS

The committee's understanding of decision analysis and its potential benefits evolved over the course of the test case. The following is a summary of the test case and some important general insights from the project.

Test Case

The problem in the test case was to find a cost-effective dredging strategy for hot spot polychlorinated biphenyl (PCB) contamination in a harbor. Four types of information were needed to run the model: (1) the set of decision alternatives (dredging volumes), (2) performance constraints or standards of PCB concentration in the tissue of edible fish, (3) the set of important factors in the decision, and (4) a decision rule for selecting the best alternative. In the test case, the objective was to select the dredged volume that minimized the total cost while meeting health and environmental objectives. There were three options: low, intermediate, and high volumes. The solution was not obvious: The low dredged volume increased the risk of not meeting the constraints (expressed as PCB concentration in the tissue of fish), whereas the high dredged volume increased the cost. Therefore, the objective was to identify the alternative that represented the best balance of risks and costs.

The basic model for the test case defines the total cost of the decision about the volume of sediment to be dredged as the sum of two factors: (1) the costs of dredging and the placement of dredged material, and (2) the cost of resource damage caused by the contaminated sediment. The data input consisted of the three dredged volumes to be evaluated, and, for each volume, the probability of not meeting the target PCB level in fish tissue (i.e., the exceedance probability), the dredging and dredged material placement cost, and the resource damage cost. All of these variables are uncertain. Uncertainty probabilities were assigned to all possible values for each variable. Detailed descriptions of the procedures and mathematical calculations involved in running the test case model can be found in Part III (below).

The results indicated that the intermediate dredged volume was preferred, followed by the high dredged volume. However, the estimate for the latter was more reliable, perhaps making this choice attractive to a risk-averse decision maker. The results are sensitive to uncertainties in both dredged volume and fish tissue concentration, because these factors strongly affect dredging costs. Interestingly, however, the results are insensitive to changes in (or disagreements about) annual resource damage costs. This counterintuitive outcome, which is discussed further in Part III, suggests how decision analysis can be used to foster understanding and consensus. In this case, the model demonstrated that differing views on the value of an important and often controversial variable should not prevent stakeholders from agreeing on the key decision.

Model Simplicity

The test case demonstrated that a decision support model need not—and perhaps should not—be comprehensive in depth or breadth. The importance of simplicity becomes apparent in the model-building exercise. The uninitiated often notice immediately what is missing from a simplified model, but the desire to "model reality" precisely is a trap. The purpose of modeling is insight. Attempting to mimic complex reality is not the best way to gain insight because the model can never be complete; more can always be added, and in the process, modeling can become an end in itself. A simple model forces the modeler to be resourceful and to think about the problem.

Another benefit of simplicity is clarity of insight. If a model is conceptually complex, then, even if it is accurate, important insights may be buried among a number of irrelevant results. Computational complexity is not a problem; calculations can be made quickly with computers. But a computer cannot overcome a confused conceptualization. If the concept is inaccurate, then errors in results may be obscured. Therefore, the most useful decision models are designed to answer specific questions, not to solve broad problems. The best models are only as complex as necessary to answer the question. Formulating specific questions, and constructing decision models specifically to answer those questions, demands careful analysis and focused discussion about the broader management problem. The disciplined thinking required to distill a complex management problem into a concise decision model provides a wealth of insight that may not end up in the model but is available to the decision maker nonetheless. In addition, this type of model building reaches closure, so attention can be shifted from model building to analysis.

Paradoxically, then, the most effective modeling strategy for solving complex management problems is to build and analyze models of simpler decisions associated with the problem and then to use both direct insights from analyzing the model and indirect insights from formulating the model to make management decisions.

Consensus Building

The test case also suggested that decision analysis—assuming the concept is understood and accepted by all parties—could be useful in bringing stakeholders together to formulate and solve problems. Without explicit models, different groups are likely to formulate the problem differently. Unintentional conceptual variations can result in unnecessary misunderstandings and distrust, which may become major obstacles to decision making. Furthermore, implicit judgments can mask how various factors interact to affect the decision outcome. The value of explicit decision models is that they provide valuable insight into the factors that drive the outcome and value of a decision alternative and how these factors interact. Decision models can assimilate expertise from a wide range of disciplines and individuals to provide a more informed analysis of the decision than an individual could manage alone. This is not to say that decision analysis can replace conventional consensus-building techniques, but it can be a valuable adjunct under certain circumstances.

Summary

Sound management of contaminated sediments can be hindered by conflicts and indecision related to the complexity, uncertainty, and volatility of the issues. To help overcome these barriers, reliable tools are needed for balancing and communicating information about risks, costs, and benefits. Decision analysis is such a tool. It can integrate multiple variables in an explicit, rigorous, and reproducible manner, and it can accommodate uncertainty. Decision analytic methods do not provide absolute solutions but they do provide insights that can be used to make balanced, well-informed decisions about the management of contaminated sediments. Decision analysis may also help foster consensus and communication among stakeholders.

There are clear advantages to collecting and analyzing relevant information in a format that can be understood and weighed by decision makers and other interested parties. Decision models can lay out the technical, scientific, and regulatory bases for decisions. In addition to fostering the sound management of contaminated sediments, decision analysis might also be used to improve government regulation. The models could help determine which regulations are controlling in a particular circumstance, thereby helping to focus the search for solutions. Even outcomes that were demonstrated to be infeasible in terms of regulatory compliance could be valuable for adjusting and, perhaps, streamlining requirements. As the federal government moves to tighten requirements for impact assessments, the use of powerful analytic tools, and the concomitant development of consistent methods of reporting data, may become increasingly attractive.

PART III: TEST CASE

The test case was developed to demonstrate the use of decision modeling and analysis in the management of contaminated sediments. Actual field data were used to ensure that the analysis was realistic. This section describes the test case in detail. The first section provides a conceptual overview of the mechanics of decision modeling. Readers unfamiliar with decision analysis or who are interested in its theoretical and conceptual underpinnings may find this information helpful. The second section describes the problem and the methodology, including the mathematical formulas used and the values chosen as parameters for the model. The third section summarizes the most important results of the decision modeling exercise.

Mechanics of Decision Modeling

Decision analysis is a computational technique for predicting the outcomes of specific candidate management approaches. Decision analysis does not provide absolute solutions but can offer valuable insights. When applied to situations involving contaminated sediments, decision analysis can be used to integrate key assessments (e.g., site characterization, risk assessment, technical feasibility studies, and economic assessment) into explicit "decision models" that describe the management problem as it appears from the perspectives of different stakeholders. The modeling approach allows stakeholders to explore disagreements about subjective elements of the problem, thereby expediting problem solving. The process also formally accounts for uncertainties.

The first steps in decision modeling are gathering information about the decision problem and selecting a set of alternatives to be evaluated with the decision model. The committee used a computerized decision model, which evaluated all possible outcomes to each candidate strategy. The possible outcomes were then rated on a value scale, which provided a measure for identifying the "best" decision alternative based on the preferences encoded in the model. The value scale can simply rate the possible outcomes from best to worst (an ordinal scale) or can quantify each possible outcome (a cardinal scale) by, for example, assigning dollar values. The preferences encoded in the analysis can be those of the decision maker or some other stakeholder.

When the possible outcomes are rated on a cardinal scale, evaluation typically involves calculating the weighted average of the values of the possible outcomes to a given alternative; weighting is based on the probability of occurrence. The weighted average is called the expected value of the alternative. The alternatives are then ranked by their expected values, and the alternative with the highest expected value is recommended.

The term "utilities" is used to describe a particular set of values that represents a decision maker's preferences among outcomes. Assuming that a set of cardinal values assigned to possible outcomes represents these preferences accurately, the axioms for the expected utility criterion for decision making—which calls for selecting the outcome that provides, on average, the highest utility—may be found in utility theory, a general theory of decision making under uncertainty (Watson and Buede, 1987).

Modern utility theory, which originated with the publication of *The Theory of Games and Economic Behavior* (von Neumann and Morgenstern, 1944), provides a normative approach for making rational decisions under uncertainty. Watson and Buede (1987) cite Savage (1954) as the most comprehensive set of principles of the expected utility criterion for rational decision making. Baumol (1972) provides a mathematical proof demonstrating that, if one wishes to act in accordance with a set of five fairly innocuous behavioral postulates (which constitute the axioms of utility theory),[4] then one should always make decisions that maximize expected utility. In practice, utilities cannot be completely known, so an "expected value maximization" criterion is often used instead of expected utility maximization. In such cases, it is important that the ranking of alternatives provided by the model be checked against judgment and intuition.

Decision models calculate the expected value of the outcome for each decision alternative (i.e., the average of all the possible outcomes to the decision alternative) and identify the alternative with the greatest expected value (i.e., the alternative for which the average of all the possible outcomes rates highest on the value scale). The part of the model that rates the possible outcomes is called the value function.

The terms in the value function are the factors that are thought to exert a significant influence on the desirability of possible outcomes. In a decision about managing contaminated sediments, these factors include, for example, the costs of dredging, sediment placement, and natural resource damage. Each factor in a value function is estimated from field data, professional judgment, and models. The level of effort, method, and sources of information used to estimate the factors in the value function must be determined on a case-by-case basis, based on the availability and reliability of possible sources of information, the preferences of the decision maker, and the total level of effort to be spent on the decision.

In some cases, the decision maker may prefer something other than the recommended (maximum expected value) alternative. This would imply that the values used to describe preferences among outcomes differ to some degree from the decision maker's utilities on the outcomes. This might happen, for example, if the outcome of the recommended alternative is less certain than the outcome of another alternative with an acceptable expected value, or if the decision maker feels that an important factor is inadequately represented in the model. In these cases, the decision maker's preferences should override the results of the decision analysis (assuming the results are properly understood). It may be helpful for the decision maker to articulate the reasons for overriding the results to ensure understanding on the part of both the decision maker and others evaluating the decision.

The effectiveness of decision analysis depends on the skill of the analyst, the effort by the decision maker to use the model and understand its results, and the quality of the information put into the model. Sometimes modeling results are counterintuitive. These are the most important results because they provide unique opportunities to learn about the problem. Counterintuitive results that hold up to scrutiny become important insights. Some counterintuitive results will not hold up, in which case the decision maker can and should override the model's recommendation (and correct the model if it is to be used again).

---

[4] The axioms of utility theory may be found in Watson and Buede (1987), page 40.

Test Case Problem

The problem in the test case is to find a cost-effective dredging strategy for a PCB hot spot in a four-square-kilometer (km^2) (roughly 1,000-acre) marine harbor.[5] The average total PCB concentration in the sediments is estimated to be 360 micrograms per gram (μg/g of carbon (C). The primary source of PCB contamination is historical discharges, which have been controlled. The placement strategy for the materials to be dredged has been selected. The present analysis focuses on how much sediment should be dredged to minimize cost while meeting health and environmental objectives for the harbor. Clear, well-defined statements of purpose like these are essential to successful decision making, especially if decision analytic tools are used.

The harbor is home to a large commercial fishing port. Although the port remains active, the harbor itself, which was formerly a commercial fishery, has been closed to fishing because of PCB contamination. Winter flounder is used as an indicator species for PCB contamination because it is a bottom-dwelling fish caught by both commercial and recreational fishermen. Total resource damage costs associated with PCB contamination in the harbor have been estimated on the order of $100 million to $1 trillion (Clites et al., 1991).

The PCB contamination levels in the harbor sediments have been mapped. Localized concentrations range from background concentration (the value is unimportant for purposes of the demonstration) to 10^5 μg PCB/g C. This wide variation makes it difficult to assess the extent of dredging required to attain a target PCB concentration level. Sampling profiles suggest that half of the PCB sediment contamination is contained in the top meter of a five-acre hot spot. Sediment PCB data suggest that, after any amount of dredging has been completed, dredging the top meter of the next-most-contaminated five acres will remove approximately half of the remaining PCBs. In other words, dredging the most-contaminated 10 acres of sediment will remove an estimated three quarters of the total PCBs, dredging the most-contaminated 15 acres will remove about 88 percent, and so on (see Figure E-1). An error in the assumed relationship between dredged volume and PCB sediment concentration may affect the decision, as is discussed in the forthcoming analysis of the test case.

Overview of Model Development

Types of Information Needed to Run the Model

For the test case, the types of information needed to run the model are (1) the specific set of decision alternatives (in this case, dredging volumes) for which the

---

[5] The case study draws on previous work by Dakins et al. (1994, 1996) and Connolly (1991).

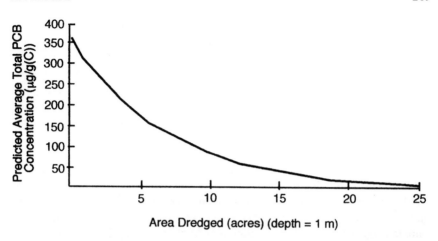

FIGURE E-1 Predicted average PCB concentration as a function of area dredged (to a depth of 1 meter), assuming sediments are dredged in order of decreasing PCB concentration.

model is to be run; (2) formal constraints on how well (at a minimum) the decision alternatives must perform (performance standards) or how much (at a maximum) must be done (stopping rules); (3) the set of important factors in the decision; and (4) a decision rule for selecting the best alternative.

Alternatives. The model has been constructed to evaluate three alternatives: low, intermediate, and maximum dredged volume. If the low or intermediate dredged volume is selected and it turns out, after dredging is completed, that the constraint (described in the forthcoming section) has not been met, then additional dredging will be done and the total volume dredged will equal the maximum (the third alternative). All three alternatives include a stopping rule, which states that dredged volume will not exceed the volume specified in the maximum dredged-volume alternative.

The first inputs into the decision model are the quantities of material to be dredged under the three alternatives (in cubic meters). It is assumed that the materials will be dredged in order of decreasing PCB concentration, according to existing maps of PCB concentrations in the harbor. Therefore, dredging locations need not be specified; they are implied in the assumption. Errors in PCB concentration maps contribute to the uncertainty about the accuracy of the assumed relationship (depicted in Figure E-1) between dredged volume and average PCB sediment concentration.

In the model's mathematical calculations, the best of the three alternatives is identified using the decision rule. Numerical values for the dredged-volume alternatives can be changed and the model run again, and the results can be compared across model runs. The total number of alternatives considered can be as many or as few as the decision maker wishes. In general, for the sake of clarity, decision models should evaluate only a few alternatives per run; additional alternatives can be explored in reruns with modified values. This strategy is more computationally efficient than analyzing many decision alternatives in a single run. Small, repeated sessions also force the modeler and decision maker to interpret results at intermediate steps in the analysis, thereby encouraging insight into the decision.

Constraints. The Food and Drug Administration (FDA) action level (the value at which the FDA requires remedial action) for total PCB concentration in edible fish tissue (2 µg/g) is the benchmark for assessing whether health and environmental objectives have been met. The measurement end-point is the sample average total PCB concentration in edible tissues of fish (flounder in this case) taken from the harbor. The constraint on the dredged-volume decision is that enough PCB-contaminated sediments must be removed to reduce the average PCB body burden to 2 µg/g of total tissue weight.

Constraints that set minimum performance standards are not a fundamental requirement of decision models. The need for minimum performance standards is established by the details of the problem at hand. However, decision models for problems in which regulatory compliance is an issue—including virtually all problems involving the management of contaminated sediment—nearly always involve minimum performance standards, which are the regulatory criteria.

The stopping rule in the test case states that dredged volume will not exceed the volume specified in the maximum dredged-volume alternative. Sensitivity of the decision to this stopping rule is examined later in this appendix. It has been argued that stopping rules are an important element of the risk management framework (Clark, 1980). Stopping rules place a tangible burden of proof on those who claim a problem exists, thereby protecting the rights of those identified as responsible for the problem. Regardless of whether stopping rules are used in real situations, they are necessary in decision modeling because they put an upper limit on the costs associated with the decision alternatives.

Factors in the Decision. There are different ways to break out the important risks, costs, and benefits affecting the selection of a decision alternative. In this case, two factors were used: the cost of dredging and dredged material placement, and the cost of resource damage. The total cost of the dredged-volume decision is modeled as the sum of the two factors. Both depend on the amount of material dredged, and both are uncertain.

Both sets of costs rise if the amount of sediment dredged is too low to meet the flounder body burden constraint because the need to perform additional dredging increases the cost of dredging and dredged material placement, and the extended duration of PCB contamination drives up the cost of resource damage.

Dredging more than enough contaminated sediments to meet the constraint, however, also increases both sets of cost, because the unnecessarily high volume adds to dredging and placement costs. (Costs of resource damage do not change; once the flounder body burden constraint is met, the marginal reduction in the cost of resource damage remains zero as the volume dredged increases.)

In summary, a reduction in the volume of contaminated sediments dredged increases the risk of not meeting the constraint on allowable flounder PCB body burden, whereas an increase in the volume increases the cost of dredging. The test case does not take into account specific benefits that may differentiate the decision alternatives (e.g., the beneficial uses for the dredged materials, the benefits of deepening the harbor, or the health benefits of minimizing the volume of contaminated materials to be disposed). The objective is simply to identify the alternative that balances the costs of dredging, placement, and resource damage in the dredged-volume decision.

Decision Rule. The decision rule in the test case was to select the alternative that minimizes the expected total cost, where the total cost is the sum of the costs of dredging and placement and resource damage. The total cost associated with each decision is uncertain. The expected total cost of an alternative is the weighted average of the possible values for its total cost, where the weighting factors are the probabilities of occurrence of each possible value.[6]

The four elements just described constitute the basic structure of the decision model. The dredged-volume decision affects the probability of meeting (or not exceeding) the flounder body burden constraint. The decision also affects both dredging and placement costs and resource damage costs. Both of these cost factors also depend on whether the flounder body burden constraint is met. The total cost is the sum of the costs of dredging and dredged material placement and resource damage.

The following information is required to run the model: the decision alternatives (dredged volumes) to be evaluated; and, for each decision alternative, the probability of not meeting the flounder body burden constraint (the exceedance probability), the cost of dredging and dredged material placement, and the cost of resource damage.

Choosing Parameters for the Model

There are many ways to provide the specific information (parameters) required to use the model. The preferred method is to use actual data on cost factors

---

[6] As a simple illustration, the expected total cost for an alternative with possible total cost values of $100,000, $200,000, and $1 million, with probabilities of 0.4, 0.5, and 0.1, respectively, is $240,000 (0.4 x 100,000 + 0.5 x 200,000 + 0.1 x 1 million).

and exceedance probabilities—if they exist. If the data are available, then the decision is straightforward and decision modeling is unnecessary. If the data are not available or are unreliable, then there are two possible strategies:

- expand the model to predict the cost factors and exceedance probability, using whatever relevant information can be found and brought to bear on the problem
- try different values for the exceedance probability and cost factors to see whether the most cost-effective decision alternative changes (an approach known as a sensitivity analysis). If the decision outcome is insensitive to the changes in these values, then it is not necessary to quantify further the cost factors or exceedance probability.

In a typical decision analysis, establishing parameters for the model involves a combination of direct data, predictive models, and sensitivity analyses. The information usually comes from a variety of sources. For example, dredging and placement cost data might be provided by a team of engineers, resource damage cost estimates by resource economists and local residents, and exceedance probability by a PCB bioaccumulation model.

Test Case Decision Model

The dredge and placement decision model developed for the demonstration project is shown in Figure E-2. This is known as an influence diagram, which is similar to a decision tree but is expressed in compact notation. Each node in the model represents either a variable or a decision. Qualitative and quantitative information about the variable or decision is contained in each node. The nodes are all annotated so the user can check the data, models, and assumptions used to establish parameters of the decision model.

The diagram contains three types of nodes: decision nodes, shown in Figure E-2 as rectangular boxes with sharp corners; value nodes, shown as rectangular boxes with rounded corners; and chance nodes, shown as ovals. Each type of node is discussed below.

The rectangular node (*dredged volume*) is a decision node and contains a list of the alternatives to be evaluated. In this case, the model evaluates three decision alternatives: low, intermediate, and maximum *dredged volume*. The alternatives are quantified by the assignment of numerical values. For the baseline analysis, the dredged volumes were set at 25,000 cubic meters (m^3), 35,000 m^3, and 50,000 m^3. The dredge depth was set at 1 meter.

Each value node (rounded rectangle) contains a set of values for the variable referred to in the node label; there is one value for each possible combination of influences. For example, the variable *resource damage cost* is influenced by two variables, *annual resource damage cost* and *length of closure*, each of which has

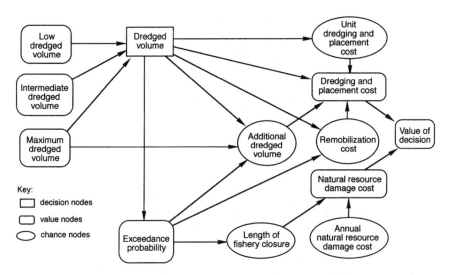

FIGURE E-2 Influence diagram of a test case. Dredging and placement model for hot spot remediation at a commercial fishery.

a set of defined states. In this model, three states were defined for *annual resource damage cost*: low, intermediate, or high. Four states were defined for *length of closure*: none, four years, five years, or six years. Therefore, *resource damage cost* has 12 possible states, corresponding to all the possible combinations of *annual resource damage cost* and *length of closure*. A value is assigned to each of the 12 states. These values can be entered directly, or they can be calculated as a function of the influencing variables. In the test case, the 12 values of *resource damage cost* are calculated as the product of *length of closure* and *annual resource damage cost*.

A value node with no influences has a single state; for example the node *low dredged volume* has one state (i.e., one volume). The states in value nodes can be revised between model runs; for example, the value used for *low dredged volume* may be increased or decreased.

Each chance (oval) node contains a set of states and values. Unlike the value nodes, a chance node can have any number of states for each combination of influencing states. For example, the node *additional dredged volume* has two states, yes and no, for each decision alternative. This node must be in one of these states or the other, but its "correct" state is uncertain. Therefore, each state is assigned a probability. In the test case, the probability of the yes state is equal to the probability of exceeding the FDA action level for contaminant concentration in the edible tissue of fish at the site (the *exceedance probability* node), given the state of the decision node (which might be set, for example, at the *minimum dredged volume*). The probability of the no state, therefore, is 1 − *exceedance probability* (because, according to the axioms of probability theory, total probability always equals 1).

The model evaluates the three dredged-volume alternatives and identifies the lowest expected cost (maximum expected value) alternative. The value function (the rule for making the decision) is coded in the terminal node of the diagram, labeled *value of decision*. The value of each possible outcome to each alternative is calculated with the formula

$$\textit{value of outcome } i = - (D\&D \textit{ cost} + \textit{resource damage cost})_{\textit{outcome } i} \quad (1)$$

where *D&D cost* is the cost of dredging and placement of the contaminated sediments, and *resource damage cost* is the cost associated with the closure of a commercial fishery (the two factors affecting the decision in this analysis). The expected value (the average value) of each decision alternative is calculated with the formula

$$\textit{expected value of alternative } j = \Sigma [-(D\&D \textit{ cost} + \textit{resource damage} (\textit{cost})_{\textit{outcome } i} \times Pr(\textit{outcome } i)] \quad (2)$$

where *nj* is the number of possible outcomes to alternative *j*.

Figure E-2 shows the variables that directly and indirectly influence the values of *D&D cost* and *resource damage cost*. The analysis could be extended to include other factors, which do not have to be measured in monetary units. The only requirement is that there must be a rule (value function) for ranking the possible outcomes to each decision and for ranking the possible decisions. If the value of outcomes is not measured on a one-dimensional cardinal scale (e.g., dollars), then decision alternatives cannot be ranked by expected value.

Parameters in the Test Case

The previous section described the user interface for a rigorous mathematical model. This section provides details of the model, including the specific rules and data contained in the nodes, as applied to the test case.

D&D cost is a deterministic function of the total dredged volume (*dredged volume + additional dredged volume*), unit dredging and placement cost (*unit D&D cost*), and the cost of remobilizing dredging, placement, and support operations if the initial dredged volume is inadequate to meet the FDA action level (*remobilization cost*). The model defines dredging and placement cost as

$$D\&D\ cost = unit\ D\&D\ cost \times (dredged\ volume + additional\ dredged\ volume) + remobilization\ cost \qquad (3)$$

There is significant uncertainty associated with *unit D&D cost*, as indicated by the use of a chance node to model this variable in Figure E-2. The unit cost may also vary with the dredging volume; that is, the values and probabilities assigned to the various states of *unit D&D cost* may be different for each of the three dredged-volume alternatives. For demonstration purposes, *unit D&D cost* was assigned a minimum value of $\$1,000/m^3$, a most likely value of $\$1,500/m^3$, and a maximum of $\$3,500/m^3$. This distribution was used for all three dredging volumes. Generally, the projects and studies from which cost data are obtained are documented in the annotation of each node, so that users of the model can critically evaluate the unit cost estimate. (The committee did not provide this information for the test case to protect the source of the data.)[7]

The definition of the variable *additional dredged volume* is very important because it provides a stopping rule for the dredged-volume decision. The following rules were used in defining this variable:

- After dredging the dredged volume (dredging depth = 1 m), contaminant partitioning in the dredged harbor will be allowed to reach a steady state, and a sampling protocol will be implemented to determine whether the contaminant concentration in edible fish tissue (as reflected in the test case by an age-adjusted sample mean from a random sample of flounder) exceeds the FDA action level.
- If the FDA action level is exceeded, then additional hot spots will be dredged (to a depth of 1 m), in order of decreasing level of contamination up to a total dredging volume of maximum dredged volume. Thus:

$$additional\ dredged\ volume = maximum\ dredged\ volume - dredged\ volume \qquad (4)$$

Of course, if the FDA action level is not exceeded, then *additional dredged volume* equals zero.

Based on these rules, the probability of having to do additional dredging is:

$$Pr(additional\ dredged\ volume > 0) = (a)\ exceedance\ probability\ if\ dredged\ volume\ is\ low\ or\ intermediate,\ or\ (b)\ 0\ if\ dredged\ volume\ is\ the\ maximum \qquad (5)$$

---

[7] The cost distribution data used in the test case were based on actual costs for a pilot project at a real location, as well as data from other sites. The data are proprietary so the location is not identified here.

The variable *remobilization cost* was estimated to be $15 million if remobilization is necessary (that is, if the FDA action level is exceeded after the initial dredging, and the initial dredged volume is less than the maximum dredged volume). This estimate was based on rough calculations of the projected cost of research and remediation for a hot spot contamination program; it is a weak link in the model but adequate for demonstration purposes. A more reliable estimate would be based on a combination of professional judgment and cost data.

Based on the rules for determining the need for additional dredging, the probability of incurring a *remobilization cost* is determined by the following formula:

$Pr(remobilization\ cost > 0)$ = (a) *exceedance probability* if *dredged volume* is low or intermediate, or (b) 0 if *dredged volume* is the maximum   (6)

Sediment contamination data and a contaminant bioaccumulation model were used to predict the probability of exceeding the FDA action level. The committee used a PCB bioaccumulation model developed by Connolly (1991), as modified by Dakins et al. (1994, 1996). As modified, this model provides age- and class-dependent probabilistic estimates of population mean contaminant concentration (TC) in flounder as a function of sediment contaminant concentration. This concentration, in turn, is assumed to be a function of dredged volume, using a model of the spatial distribution of sediment contaminant concentration at the site. Thus:

$exceedance\ probability = Pr(TC \geq FDA\ action\ level)$
$= f(dredged\ volume)$ (7)

For the three dredged volumes selected (25,000 m^3, 35,000 m^3, and 50,000 m^3), the exceedance probabilities calculated with the food web model are 0.48, 0.25, and ~0.001, respectively. A detailed presentation of the use of the bioaccumulation model to estimate *exceedance probability* as a function of volume of sediment dredged can be found in Dakins et al. (1994).

The second factor in the value function is *resource damage cost*, which represents the total economic impact of fishery closure. This factor was estimated as the product of two probabilistic variables, *length of closure* and *annual resource damage cost*. *Length of closure* represents the elapsed time (years) between the determination that initial dredging did not meet the FDA action level and the reopening of the fishery following additional dredging. The *length of closure*, if the FDA action level were exceeded, was estimated to be four to six years. Probabilities were assigned based on professional judgment (with five years the most likely *length of closure*, and the sum of the probabilities equal to one):

Length of Closure (years)	Probability	(8)
0	1 − *exceedance probability*	
4	0.25 × *exceedance probability*	
5	0.50 × *exceedance probability*	
6	0.25 × *exceedance probability*	

Variability in *annual resource damage cost* due to disagreements about risks, costs, and benefits is likely to exceed variability due to uncertainty because stakeholders can have widely divergent opinions about the value of this factor. Because it reflects opinions instead of facts, variability due to disagreement should be treated parametrically rather than probabilistically. That is, the model should be run repeatedly using different uncertainty distributions for *annual resource damage cost*, with each run representing a different opinion about this cost. For the test case, the committee used a distribution based loosely on a single economic analysis of the total impact of the closure of a commercial fishery (Clites et al., 1991):

Annual Resource Damage Cost	Probability	(9)
$7 million	0.1	
$12 million	0.8	
$17 million	0.1	

Test Case Results

This section summarizes some of the key results of the decision modeling exercise. As noted earlier in this appendix, the most important results are insights that influence the direction of the analysis, rather than bottom-line recommendations. The use of decision modeling software[8] allowed the committee to construct and display decision models as diagrams and spreadsheets, rather than as lines of computer code, and to display model output as graphs and diagrams. The software also made it possible to perform valuable exploratory analyses of the decision problem in ways that otherwise would have been impractical.

The set of parameter values given in the preceding section was used as the baseline for the dredged-volume decision model. All the nodes in the model were annotated with information about the data and assumptions used to develop the baseline. These values and assumptions were altered during the analysis to test

---

[8] The committee identified four commercial decision modeling software packages: DATA (TreeAge), Demos (Lumina Decision Systems, Inc.), DPL (Applied Decision Analysis, Inc.), and Supertree (Decision Education Center). This analysis was performed using DPL.

assumptions and explore the sensitivity of the results to various changes. The set of decision alternatives (the numbers of cubic meters in each alternative) was also changed to determine the optimal dredged volume for the baseline values.

Baseline Results

Figure E-3 show the results of the analysis for the three initial decision alternatives, using the baseline parameter values. The display is a simplified decision tree showing the expected value of each alternative. Using the criterion of highest expected value, the preferred alternative is the intermediate dredged volume (35,000 m^3), with an expected value of −$96.25 million, followed by the alternative of immediately dredging the maximum dredged volume (50,000 m^3), with an expected value of −$100 million.

Figure E-4 provides more information about these results—information that would not be readily visible without the graphic representations. The three curves are cumulative probability distribution functions for the values of the alternatives. The cumulative probability distribution functions for the maximum dredged volume is the steepest (most vertical) curve; the range of possible values along the horizontal axis (from −$150 million to −$75 million) is much narrower than for the other two cumulative probability distribution functions. This narrow range

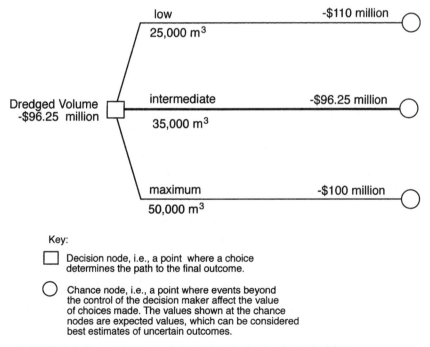

FIGURE E-3 Expected values of alternative dredged-volume decisions.

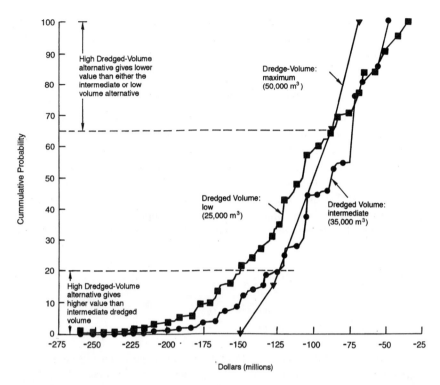

FIGURE E-4 Dredged-volume decision analysis.

means that the value of the maximum dredged-volume alternative is more certain than the values for the other two alternatives. (In this context, value is the negative of cost; a value of –$150 million is equivalent to a cost of +$150 million.)

The relatively low level of uncertainty associated with the maximum dredged-volume alternative makes sense because the low and intermediate dredged-volume alternatives involve more uncertainty about the remobilization and resource damage costs. The greater reliability (i.e., lower uncertainty) of the maximum dredged-volume scenario could persuade a risk-averse decision maker to select this alternative, despite its higher expected cost.

Another way of interpreting Figure E-4 is to note where the cumulative probability distribution functions intersect. The high and intermediate dredged-volume alternatives intersect at a cumulative probability level of approximately 20 percent (shown on the vertical axis), meaning that the model predicts about a 20 percent chance of the high dredged volume providing a better outcome than the intermediate dredged volume. However, the point where the high and low dredged-volume alternatives intersect predicts a 35 percent chance that the high dredged-volume alternative will have the worst outcome of the three alternatives considered.

This exercise indicates how decision modeling, and the visual display of the results, can foster understanding of the problem in ways that numerical formulas alone cannot.

Analysis of Alternative Scenarios

The values used as model parameters were changed to test assumptions and explore the sensitivity of the results. For example, the low uncertainty for the maximum dredged-volume alternative may hinge on the decision rule stating that the dredged volume will not exceed the defined maximum of 50,000 m^3. On the other hand, the exceedance probability for a dredged volume of 50,000 m^3 may be low enough that the uncertainty about this alternative is insensitive to the upper limit on the dredged volume. To test this assumption, the model was modified. The three decision alternatives remained the same, but the additional dredged volume required if the FDA action level were exceeded was defined as:

Additional Dredged Volume	Probability	(10)
50,000 m^3 – *dredged volume*	0.5	
70,000 m^3 – *dredged volume*	0.5	

Figures E-5 and E-6 show the effect of the added uncertainty about the maximum dredged volume. Figure E-5 shows that the expected values of the low and intermediate dredged volumes have dropped compared to the baseline (shown in Figure E-3), whereas the value of the maximum dredged volume is essentially unchanged. By comparing Figure E-3 with Figure E-5, it can be seen that a 50 percent chance of having to dredge up to 70,000 m^3 (see Equation 10) has essentially no effect on the outcome of the highest dredged-volume alternative. (The probability of exceeding the FDA action level is low [on the order of 0.001] if the initial decision is to dredge 50,000 m^3.) On the other hand, the exceedance probabilities for the intermediate and low dredged-volume alternatives are sufficiently high that the added uncertainty about a possible penalty (assessed if the initial dredged volume is too low) drives down their expected values.

Another issue of interest to the committee was whether and how decision modeling could help resolve disagreements about values, such as the local economic impact of a sediment management decision. As an illustration, there might be significant disagreement in the test case about the *annual resource damage cost* of the closure of a fishery. For example, if a substitute existed for the damaged fishery, then one might argue that the resource damage should be set at the marginal cost of the closure, rather than the total cost. The impact of such disagreements on the decision outcome can be analyzed using a technique known as switchover, or policy region, analysis (Morgan and Henrion, 1990). In a switchover analysis, the model is run repeatedly, each time using a different

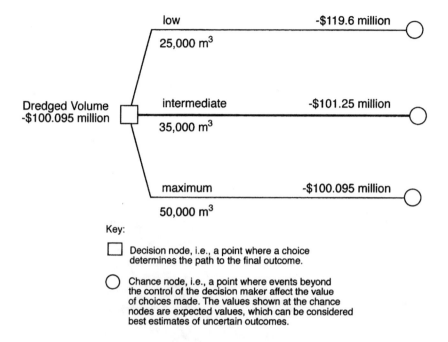

FIGURE E-5 Expected values of alternative dredged-volume decisions with modified model parameters.

estimate for the variable being analyzed. The computer determines the values at which the model's preference "switches" from one alternative to another.

Figure E-7 shows the results of a switchover analysis for *annual resource damage cost*. The objective is to examine the sensitivity of the preferred decision alternative to uncertainty and/or disagreement about model parameter values. In this case, the value for this cost was varied from $0 to $20 million to determine the effect on the *preferred decision* (where the *preferred decision* is defined as having the highest expected value). All other model values and probabilities remained the same as in the baseline analysis.

The switchover analysis shows that the preferred decision (intermediate dredged volume) is insensitive to *annual resource damage cost* over a wide range of values (from $43,479 to $15 million). These limits were determined using the computer to perform repeated runs of the model. When the cost drops below $43,480, the low dredged volume becomes the preferred alternative (because the risk associated with the exceedance probability for this alternative diminishes). When *annual resource damage cost* exceeds $15 million, the maximum dredged volume becomes the preferred alternative (because as this cost rises, the risk associated with the exceedance probability for the intermediate dredged volume decision becomes unacceptably high).

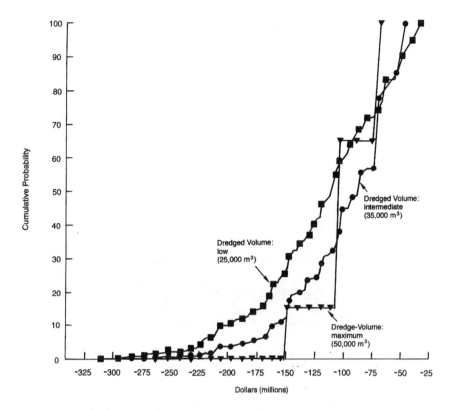

FIGURE E-6 Effect of modified model parameters on maximum dredged volume decision.

This is an enlightening exercise because it shows that disagreement about the economic impact of closure of the fishery does not have much impact on the decision, at least under the conditions and assumptions in the baseline model. If this conclusion held up over a range of variations to the model, then disputes concerning the actual value of resource damage could be set aside. And, if variability in cost estimates were due to disagreements (rather than uncertainties) about risks, costs, and benefits, then expensive analyses, which probably would not resolve the question anyway, could be avoided.

Summary of Results

The analysis of deciding how much sediment to dredge in the test case indicates that the low dredged-volume alternative probably can be eliminated from further consideration because it is consistently outperformed by the other two alternatives. The analysis also clarifies the issues involved in choosing between the intermediate and high dredged volumes. The high dredged volume would be

APPENDIX E 283

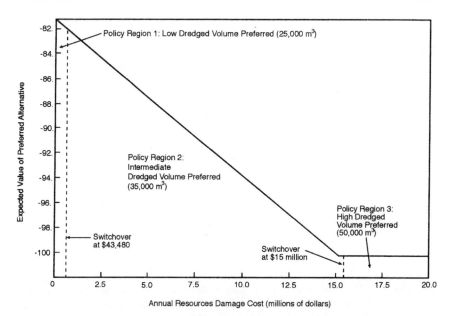

FIGURE E-7 Switchover analysis for the *annual resource damage cost*.

preferred by a risk-averse decision maker because the outcome is more predictable than the intermediate dredged volume and the expected value is nearly as high.

Under the baseline scenario, the intermediate dredged volume is preferred, but the expected value is only $3.75 million (4 percent) higher than the value for the high dredged volume (as shown in Figure E-3), which has a more certain outcome. The range of possible costs for the high dredged-volume alternative is $75 million to $150 million, whereas the range of possible costs for the intermediate dredged-volume alternative is $50 million to $250 million (as shown in Figure E-4). The model predicts an approximately 20 percent chance of the high dredged volume providing a better outcome than the intermediate dredged volume. However, the model also predicts a 35 percent chance that the high dredged-volume alternative will have the worst outcome of the three alternatives.

If the model is altered slightly by increasing the uncertainty of the maximum dredged-volume alternative, then the high dredged-volume alternative becomes the model's preferred choice, demonstrating why a risk-averse decision maker might override the recommendation of the baseline analysis.

REFERENCES

Baumol, W.J. 1972. Economic Theory and Operations Analysis. Englewood Cliffs, New Jersey: Prentice-Hall.
Baird, B.F. 1989. Managerial Decisions Under Uncertainty: An Introduction to the Analysis of Decision Making. New York: John Wiley & Sons.

Carpenter, S.L., and W.J.D. Kennedy. 1988. Managing Public Disputes: A Practical Guide to Handling Conflicts and Reaching Agreements. San Francisco, California: Jossey-Bass.

Clark, W.C. 1980. Witches, Floods, and Wonder Drugs: Historical Perspectives on Risk Management in Societal Risk Assessment: How Safe is Safe Enough. R.C. Schwing and W.A. Albers, eds. New York: Plenum Press.

Clemen, R.T. 1991. Making Hard Decisions: An Introduction to Decision Analysis. Boston: Kent Publishing Company.

Clites, A.H., T.D. Fontaine, and J.R. Wells. 1991. Distributed costs of environmental contamination. Ecological Economics 3:215–229.

Connolly, J.P. 1991. Application of a food chain model to polychlorinated biphenyl contamination of the lobster and winter flounder food chains in New Bedford Harbor. Environmental Science and Technology 25:760–770.

Dakins, M.E., J.E. Toll, and M.J. Small. 1994. Risk-based environmental remediation: Decision framework and the role of uncertainty. Environmental Toxicology and Chemistry 13(12):1907–1915.

Dakins, M.E., J.E. Toll, M.J. Small, and K.P. Brand. 1996. Risk-based environmental remediation: Bayesian Monte Carlo analysis and the expected value of sample information. Risk Analysis 16(1):67–79.

Environmental Protection Agency (EPA). 1994. Guidance for the Data Quality Objectives Process. Final Report. EPA QA/G-4. Washington, D.C.: EPA Quality Assurance Management Staff. September.

EPA. 1996. Review of National Ambient Air Quality Standards for Ozone: Assessment of Scientific and Technical Information: OAQPS Staff Paper. EPA 542/R-96-007. Office of Air Quality Planning and Standards. Research Triangle Park, North Carolina: EPA.

Horton, C. 1993. To Build or Not to Build: Economic Assessment Model Helps Companies Decide Whether to Build Onsite Wastewater Treatment Facilities or Buy Services. Industrial Wastewater 1(2):39–43.

Howard, R.A., and J.E. Matheson, eds. 1984. The Principles and Applications of Decision Analysis. Menlo Park, California: Strategic Decision Group.

Keeney, R.L., and H. Raiffa. 1976. Decisions with Multiple Objectives. New York: John Wiley & Sons.

Morgan, M.G., and M. Henrion. 1990. Uncertainty: A Guide to Dealing with Uncertainty in Quantitative Risk and Policy Analysis. Cambridge, United Kingdom: Cambridge University Press.

National Research Council (NRC). 1989. Contaminated Marine Sediments—Assessment and Remediation. Washington D.C.: National Academy Press.

Northwest Power Planning Council (NWPPC). 1991. 1991 Northwest Conservation and Electric Power Plan, vol. 2, part 2. Portland, Oregon: NWPPC.

Raiffa, H. 1968. Decision Analysis: Introductory Lectures on Choices Under Uncertainty. Reading, Massachusetts: Addison-Wesley.

Savage, L.J. 1954. The Foundation of Statistics. New York: Wiley & Sons.

Silkien, R.L. 1990. Decision Analysis and Quantitative Risk Characterization: Information-Analysis Based Risk Characterization. Washington, D.C.: Chemical Manufacturers Association, Inc.

Singer, L.R. 1990. Settling Disputes. Boulder, Colorado: Westview Press.

U.S. Department of Energy (DOE). 1986. A Multi-Attribute Utility Analysis of Sites Nominated for Characterization for the First Radioactive-Waste Repository—A Decision-Aiding Methodology. DOE/RW-0074. Office of Civilian Radioactive Waste Management. Washington D.C.: DOE.

von Neumann, J., and O. Morgenstern. 1944. The Theory of Games and Economic Behavior. Princeton, New Jersey: Princeton University Press.

Watson, S.R., and D.M. Buede. 1987. Decision Synthesis: The Principles and Practice of Decision Analysis. Cambridge, United Kingdom: Cambridge University Press.

Index

A

Abyssal plain, 136–137
Acceptable risk, 3, 22
Accountability, 8, 25, 63, 169
Acid leaching, 121, 127
Acoustic profiling, 9, 73–74, 77, 107, 170
Arbitration, 54
Atmospheric fallout, 15

B

Bayou Bonfouca, 106, 108, 109, 111, 116
Benchmark values, 22
Beneficial uses of sediments, 7–8, 50, 56–59, 61, 118, 171–172
Bioremediation
  costs, 104
  current understanding, 102–103
  ex situ, 12, 127–130, 132, 166–167
  experience with, 100–102, 103
  recommendations, 164–165, 166–167
  research needs, 130, 165
  in situ, 11, 100–104
Biosensors, 74–75
Bioslurry reactors, 127
Bioturbation, 66

Black Rock Harbor, 56
Boston Harbor, 54, 69, 134, 227–229

C

Cadmium, 25
Capping, in-place
  advantages, 95, 163–164
  conditions for, 10–11
  contained aquatic disposal and, 134
  costs, 95
  current understanding, 94
  current utilization, 94
  definition, 94
  design, 130, 164
  effectiveness, 145
  goals, 94
  indications for, 94–95
  as interim control, 90
  limitations of, 95
  materials for, 94
  monitoring, 95, 97
  opportunities for technical
    improvement, 97
  recommendations, 164
  regulations, 11, 95, 97, 164, 190
  strategies for augmenting, 95–97

Case histories, 26, 27, 54
  decision analysis test case, 263, 268, 272–283
  sites selected for, 225. *See also specific site*
Chemical contaminants, 1
  characteristics of sediments, 23, 24
  chemicals of concern, 23
  as long-term threat, 23
  management challenges, 24
  mixture, 23–24
Chemical destruction, 125–127
Chemical sensors, 9, 74–75, 77, 170
Chemical separation, 12, 121–123
Chemical treatment, in situ, 11, 97–100, 164
Chesapeake Bay, 196
Clean Water Act (CWA), 5, 8, 18, 25, 182
  evaluation methodology, 5–6
  interrelationship with other laws, 214–220
  local/state regulations and, 209–211
  practicability concept, 47 n.3
  risk assessment, 47, 156
  sediment placement requirements, 48, 49 n.5, 187
  sediment-relevant provisions, 192–196
  source control provisions, 63–64
Coastal Zone Management Act (CZMA), 5, 57, 182, 211–212
Collaborative problem-solving, 54
Composting, 129
Comprehensive Environmental Response, Cleanup, and Liability Act. *See* Superfund
Confined disposal facilities (CDFs)
  advantages of, 11
  bioremediation in, 127–129
  chemical contaminants in, 24
  contaminant migration pathways, 132
  costs, 132
  design features, 130
  disadvantages of, 11
  goals, 130
  for rapid response, 90
  recommendations, 165
  recovery/reuse, 134
  research needs, 141, 165
  technological augmentation, 132
  treatment strategies in, 132
  use of, 9
Conflict resolution, 54–55, 259
Contained aquatic disposal (CAD), 12, 134–137, 141, 166
Control of sediments
  comparative analysis of technologies, 12–13, 142–147
  definition, 2 n.2, 16 n.1
Cooperative research and development agreements, 58 n.8
Corps of Engineers, U.S. Army (USACE), 5
  authority and responsibilities, 18–19, 25
  beneficial uses of sediments, 57–58, 61
  contained aquatic disposal guidelines, 134–136
  cost-benefit analysis in, 40–41, 252–253
  decision making framework, 30–32
  dispute resolution policy, 55
  dredged material evaluations, 48
  dredged material management plans, 52
  dredging permits, 184–185
  local/state regulations and, 210–211, 213
  recommendations for, 161, 167–168, 171, 172
  WRDA provisions, 199–200, 201–202, 204–208
Cost-benefit analysis
  constrained port capacity in calculations of, 252–253
  current application, 4, 51, 155–156
  environmental costs in, 253–255
  expanded application, 4, 40–41, 42
  principles of, 3–4, 240–241
  process, 37–39, 241–243, 245–246
  rationale for new dredging projects, 21
  recommended utilization, 157, 160, 161
  role of, 36–37, 39–40, 239–240, 243–244, 255
  sediment disposal regulation, 6–7
  types of benefits/costs, 37, 246–252
Cost effectiveness
  data needs for analysis of, 140–141
  definition, 3 n.3, 27 n.9

INDEX
287

dredging, 11
  management goals, 27–28
  political challenges, 26–27
  regulatory obstacles, 47, 60
  site characterization for, 8–9, 10, 77
Cost of management
  accountability of polluters, 8, 25
  allocation, regulatory requirements for, 50–51, 157, 197–198
  beneficial uses of sediments, 57
  bioremediation, 104
  chemical immobilization, 123
  comparison of remediation strategies, 10, 12–13, 140, 162
  confined disposal facilities, 132, 134
  current estimates, 10
  data needs for technology comparison, 162–163
  determinants of, 10, 20
  dredging costs, 105
  engineering costs of cleanup, 162
  ex situ bioremediation, 127
  ex situ management, 116
  in-place capping, 95
  interim controls, 88
  landfill disposal, 118, 132
  metal leaching, 121
  natural recovery, 92
  navigational dredging, 20
  new-work dredging, 50
  particle separation, 140
  physical separation, 120–121
  rationale for navigational dredging, 20–21
  separation of organic contaminants, 121–123
  sharing, 8
  soil washing, 134
  spectrum, 27
  Superfund site cleanup, 27
  thermal desorption, 123
  thermal destruction, 125
  trade-offs, 27–28, 34–35
  WRDA provisions on allocation, 197–198
CWA. *See* Clean Water Act
CZMA. *See* Coastal Zone Management Act

D

Data collection/management
  for cost-effectiveness analysis, 140–141
  cost information, for technology comparison, 162–163
  for evaluation of site dynamics, 70–71
  field surveys, 71–73, 77
  goals of detailed site assessment, 33, 73
  identifying goals for, 32–33
  local transportation departments as sources, 70–71
  monitoring in-place caps, 97
  for numerical modeling of sediment transport, 76
  recommendations for, 168
  site history, 67–70
Decision analysis
  benefits of, 260, 265
  choosing parameters for, 271–272
  for consensus building, 264–265
  data needs, 268–271
  dispute resolution and, 259–260
  mechanics of modeling, 266–267
  model simplicity, 264
  potential applications, 4, 41, 42, 258, 261
  principles of, 4, 258
  recommended utilization, 160, 161
  risk balancing in, 260–261
  role of, 41, 54, 258
  state of practice, 261–262
  test case, 263, 268, 272–283
Decision making
  agency frameworks for, 30–32
  consensus building, 54–55, 158, 161
  cost-benefit analysis for, 36–41, 239–240, 255
  cost information for, 140–141
  decision analysis for, 41
  goals, 28
  identifying decision criteria, 32
  improving project implementation, 168–172
  for near-term improvements in sediment management, 154
  obstacles to, 25–26

obstacles to effectiveness in, 154–155
opportunities for improvement, 155
phases of stakeholder involvement, 53
political context, 7
project-specific considerations, 45, 62
recommendations for improving, 155–161
regulatory system influence, 6–7, 155–156
risk communication for, 56
selection of interim controls, 87–88
significance of regulatory framework, 45
significance of stakeholder interests, 45
site sampling design, 72
stakeholder interests, 7–8
technology assessment for, 45–46, 84
timeliness of, 5, 48–50
tools for, 2–4, 35, 257
trade-offs in, 13, 34–35, 37, 147
value-driven factors, 45
Detroit River, 73
Dioxin, 23
in Newark Bay, 25
Discharged sediments, 16 n.2
Disposal of sediments
in abyssal plain, 136–137
beneficial uses, 7–8, 50, 56–59, 61, 118, 171–172
in confined disposal facilities, 130–134
contained aquatic disposal, 134–137
cost of, 6–7
inadequate risk analysis in decision making, 156–157
inconsistent regulatory system, 6–7, 46–48
in landfills, 132, 137
permitting process, 6, 47–48
regulatory system, 25, 185–187
safety guarantees, 59
shortage of placement space, 51–52
terminology, 16 n.2
Dispute resolution, 55, 259–260
Distribution of contaminants (aquatic process), 1, 33, 64–67
bioturbation, 66
contaminant resuspension in dredging, 109–111

core sample evaluations, 72
erodibility of sediments, 66–67
evaluation of site dynamics, 70–71
field surveys, 71–73
fine-grained sediments, 65, 120
fluff layer, 65
interim control of, 89–91
mobility of sediment layers, 66
nonlinear behavior, 65
numerical modeling of processes, 75–76
organic matter degradation and, 66
particle aggregation, 65
retention sites, 71
sediment-water interface, 65, 66
site characterization, 64–67, 69
site sampling, design of, 72
understanding of marine environments, 80, 92
Dredged material management plans, 52
Dredging, environmental
cable arm clamshell dredge, 108, 114
for contained aquatic disposal, 136
contaminant release in, 109–111
contract bidding process, 104, 116
cost, 105, 140
digital system, 114–115
dry excavation for, 113
equipment for, 105–106
pneumatic barrier for, 113
precision in, 107–109, 165
recent innovations, 113–116
recommendations, 165
for removal of contaminated sediments, 104–105
research needs, 141
silt curtains for, 112
site assessment, 106, 107
site isolation for, 112
systems approach, 105
See also Dredging, navigational; Dredging technology
Dredging, navigational
applicable legislation, 18
channel maintenance, 21
cost, 10, 20, 105
cost allocation, 50–51, 63

cost-benefit analysis, 51
disposal regulation, 6–7
economic rationale, 20–21
federal oversight agencies, 5
governing legislation, 5
interagency collaboration for
  permitting, 49
new construction, 21, 50 n.6
regulatory system, 5, 184–187
sediment traps in, 90–91
source control strategies, 63
as source of contaminated sediment, 19
trade-offs in decision making, 34–35
volume of sediments removed, 20, 21, 187
See also Dredging, environmental;
  Dredging technology
Dredging technology
  backhoe dredges, 106
  bottom-crawling systems, 108, 114
  cable arm clamshell, 108, 114
  depth of cut control, 108–109
  digital system, 114–115
  hopper dredges, 105–106, 111
  hydraulic equipment, 105, 111
  mechanical equipment, 105, 111
  positioning systems, 107–108
  precision technologies, 11, 107–109, 140, 165
  See also Dredging, environmental;
    Dredging, navigational
Duwamish Waterway, 134

E

Ecosystem functioning, 15–16
End-points, 22, 36, 82
Endangered Species Act, 182
Energy, U.S. Department of, 263
Environmental cleanup
  applicable legislation, 16–18
  engineering costs, 162
  extent of need for, 19
  motivation for, 19
  treatment strategies, 18
  See also Dredging, environmental;
    Superfund

Environmental impact statement, 184–185
Environmental Protection Agency (EPA), 5, 49, 200
  Assessment and Remediation of
    Contaminated Sediments program
    (ARCS), 81
  decision making framework, 30–32
  dispute resolution policy, 55
  dredged material evaluations, 48
  dredging permits, 185–186
  recommendations for, 161, 167–168, 172
  responsibilities, 18
  risk assessment paradigm, 36 n.2
  sediment quality criteria, 64
  survey of sediment quality, 204–204
EPA. *See* Environmental Protection
  Agency
Evaluation of technologies
  comparative, 12–13, 142–147
  cost-effectiveness, 140–141
  methodology for, 84
  performance monitoring, 137–141
  research needs for, 141–142
Ex situ management, 84
  biological treatment, 127–130, 166–167
  chemical immobilization, 123
  chemical separation, 121–123
  comparative analysis of remediation
    technologies, 12–13, 146–147
  confined disposal facilities, 130–134, 165
  contained aquatic disposal, 134–137, 166
  containment strategies, 130, 147
  cost, 116, 140
  cost of biological treatment, 127
  cost of chemical immobilization, 123
  cost of metal leaching, 121
  cost of nucleophilic substitution, 127
  cost of physical separation, 120–121
  cost of separating organic contaminants, 121–123
  cost of thermal desorption, 123
  cost of thermal destruction, 125
  current status of strategies, 10, 12, 161
  geotextile containers, 136

goals, 116
indications for, 116
interim storage facilities, 116–117
landfill disposal, 137
long-term monitoring, 139
operations in, 117–118
physical separation technologies, 117, 118–121
recommendations, 165–167
research needs, 141, 166, 167
solids-water separation, 118
thermal desorption technologies, 123
thermal destruction, 123–125

F

Fiber-optic sensors, 74–75
Fish/shellfish industries
  interim controls on, 88–89
  threat of contaminated sediments, 15–16
Fluff layer, 65
Foreign trade
  significance of, in national economy, 21–22
  waterborne volume, 21
Freezing, soil-water, 100

G

Geological Survey, U.S., 70
Geotextile containers, 136
Global positioning system, 107
Great Lakes, 81, 196

H

Hamburg, Germany, 120
Harbor Maintenance Trust Fund, 21
Hart and Miller islands, 54, 56, 229–231
Health advisory, 88–89
Heavy metals. *See* Metals/heavy metals
Hot spots, 11, 15, 28
  interim technological intervention, 90
  regulatory provisions, 191
Housatonic River, 103

Hudson River, 89, 103, 196
Hydrocyclones, 120

I

Implementation
  delays in, 25–26
  recommendations, 168–172
  stakeholder interests, 7–8
In situ management
  biodegradation, 11, 164–165
  biological treatment, 100–104
  chemical immobilization, 97–99
  chemical sensors for, 74–75
  chemical treatment, 11, 99–100, 164
  comparative analysis of technologies, 12–13, 145–146
  cost, 140
  disadvantages, 10, 163
  freezing, 100
  goals, 91
  long-term monitoring, 139
  natural recovery, 91–92, 94, 163
  recommendations, 163
  research needs, 141
  treatment strategies, 97
  types of, 91
  utilization, 91
  See also Capping, in-place
Incineration, 118, 123–125, 139
Indiana Harbor, 115
Inland waterways, sediment disposal in, 25
Innovation
  in dredging technologies, 114–116
  impediments to, 28
  recommendations for research and development, 167–168
  research and development process for, 141–142
  in site assessment, 73–75
Interim controls, 9, 33, 82
  administrative, 88–89, 145
  comparative analysis of technology categories, 12–13, 145
  compatibility with long-term strategy, 88

cost considerations, 88
definition, 87
effective use, 87
effectiveness of, 170–171
indications for, 82, 87
monitoring effectiveness of, 89–90, 139
role of, 86–87
selection of, 87–88
technology-based, 89–91, 145
types of, 87
International agreements, 47, 94
International Convention on the Prevention of Marine Pollution by Dumping of Wastes. *See* London Convention of 1972

J

James River, 88, 91, 145, 231–233

L

Lake Hartwell, 91
Land placement, 25, 118
advantages/disadvantages, 137
for aerobic degradation, 127
cost allocation, 50
cost of disposal in, 132
regulation, 185
sediment handling for, 137
Laser positioning systems, 107–108
London Convention of 1972, 47, 94
Long Island Sound, 196
Long-term considerations
chemical contaminants, 23
compatibility of interim controls, 88
management of sediments, 10–12, 33–34
monitoring, 139
technology recommendations, 167–168

M

Management of sediments
challenges, 1, 16, 27–28, 44
conceptual overview, 30–34
definition, 2 n.2, 16 n.1

empirical knowledge base, 81
interim controls, 9, 33, 82, 86–91, 139, 145, 170–171
long-term plans/strategies, 10–12, 33–34
obstacles to innovation, 28
opportunities for near-term improvements, 154
performance evaluation, 137–141
site characterization, 8–9
site-specific considerations, 45, 62, 82
source control, 8
systems approach, 2–3, 34
themes, 44–45
types of strategies, 2 n.2, 16 n.1. *See also specific type of strategy*
understanding of marine environments, 80–81
Manistique Harbor, 90, 115, 120
Manitowoc Harbor, 99
Marathon Battery, 69, 123, 233–234
Marine Protection, Research and Sanctuaries Act (MPRSA), 5–6
disposal requirements, 46–47, 48, 185, 186
risk assessment, 155–156
Maritime Administration, 49
Mediation, 54
Mercury, 25
Metals/heavy metals, 56, 57
chemical immobilization, 123
chemical separation, 121, 127
thermal destruction, 125
Mitigation, off-site, 59–60
MPRSA. *See* Marine Protection, Research and Sanctuaries Act

N

National Ambient Air Quality Standards, 261, 262
National Dredging Team, 49, 52
National Environmental Policy Act (NEPA) of 1969, 91, 182, 184
National Ocean Survey, 70
National Oceanic and Atmospheric Administration, 5, 18

National Priorities List, 19
National Weather Service, 70
Natural recovery
 advantages, 10, 92, 163
 applications, 91
 in confined disposal facilities, 132
 cost, 92
 disadvantages, 10
 effectiveness, 10, 145
 indications for, 23, 91–92
 limitations, 92
 monitoring, 92
 recommendations, 163
 research needs, 94, 163
Negotiated rule making, 54
NEPA. *See* National Environmental Policy Act of 1969
New Bedford Harbor, 89, 90, 95, 111, 112, 123
New York, Port of, 49
Newark Bay, 25
Nucleophilic substitution, 125–127
Numerical modeling, 75–76
 of sediment resuspension in dredging, 110

O

Ocean dumping
 in abyssal plain, 136–137
 international agreements, 47
 regulatory system, 25, 185, 186–187, 199
Ocean Dumping Act. *See* Marine Protection, Research and Sanctuaries Act
Organohalogens, 25
Oxidant injection, 99–100

P

Palos Verdes slope cleanup, 114
Particle separation, 117, 118–121, 140
Permitting process
 differences among agencies, 6, 47–48
 disposal of sediments, 185
 interagency collaboration, 49

 risk assessment in, 60, 156
 time delays, 49
Pesticides, 100
Petroleum products, 25, 100
Placement
 definition, 16 n.2
 See also Disposal of sediments
Pneumatic barrier, 113
Political environment
 common concerns of stakeholders, 27
 for effective decision making, 7
 as obstacle to effective sediment management, 26–27
 obstacles to decision making, 25–26
 stakeholder interests, 7–8, 52
Polyaromatic hydrocarbons, 23, 99, 100, 104, 125, 129
Polychlorinated biphenyls (PCBs), 23, 88–89, 99, 100, 103–104, 111, 112, 120, 123, 125, 127–129, 196, 254, 263, 268
Post-project evaluations, 82, 139
Practicability, 47 n.3
Public awareness/perception
 of aquatic processes, 26–27
 citizen stakeholders, 52–53
 effectiveness of health advisories, 89
Pyrolysis, 125

Q

Quantity of contaminated sediments, 1

R

RCRA. *See* Resource Conservation and Recovery Act
Regulatory system
 barriers to capping, 11
 on beneficial uses of sediment, 56, 58, 61
 bioremediation issues, 129–130
 cost allocation in, 50–51, 60, 157
 determinants of applicability, 24, 183–184
 for disposal of sediments, 25, 156–157

INDEX 293

dispute resolution policy, 55
for environmental cleanup, 16–18
evaluation of placement alternatives, requirements for, 46–48
foreign-flag dredges, 113
gaps in, 220–221
government role in developing placement space, 51–52
hot spot management, 191
in-place capping, 95, 164
inconsistency in, 5–7, 25–26, 155–157
interpretation of legislative intent, 49–50, 60
legislative/agency interrelationships, 214–220
natural resource damage claims, 190–191
navigation-related, 184–187
obstacles to effective remediation in, 154–156
opportunities for improvement, 5–7, 48, 60–61, 156–157
potential reforms, 221–224
recommendations for improving, 155–157
reform initiatives, 49
relevant federal agencies, 5, 18
relevant legislation, 5, 183
scope of, 181–182
shortcomings of, 5, 24, 25–26, 46, 48, 60
significance of, for management of sediments, 44, 45
site cleanup legislation/oversight, 187–191
source control through, 63–64
state programs, 208–214
timeliness of decision making in, 48–50
water resource public works projects, 197
See also Permitting process
Remediation
comparative analysis of technology categories, 12–13, 142–147
conceptual management approach, 84
conceptual model, 82
cost-effectiveness analysis, 140

current utilization, 84, 161
definition, 2 n.2, 16 n.1
determinants of strategy selection, 16
empirical knowledge base, 81
evaluation of technologies, methodology for, 84
goals, 82
importance of source control in, 63
legal obstacles to effectiveness in, 154–155
long term controls, 10
recommendations for technologies, 161–168
research needs, 141–142
state of the technology considerations in decision making, 45–46
subsystem components/structure, 82, 84. *See also specific component*
Replacement habitat, 59–60
Research needs
beneficial uses of sediments, 58–59
bioremediation, 130, 165
chemical treatment in situ, 164
confined disposal facilities, 165
contained aquatic disposal, 166
in-place capping, 97
natural recovery, 94, 163
numerical modeling of sediment transport, 76
recommendations, 167–168
remediation methods, 141–142
Residual risks, 34, 35
Resource Conservation and Recovery Act (RCRA), 5, 18
sediment placement requirements, 48
Risk analysis
activities of, 3, 35
current application, 3, 35
expanded application, 3
opportunities for improvement, 36, 42
recommended utilization, 159
Risk assessment
cost-benefit analysis and, 36–37
current application, 3, 35
EPA paradigm, 36 n.2
goals, 3, 22, 35
in management strategy planning, 33

methodological differences among agencies, 6, 155–156
recommended utilization, 156–157, 159
residual risks, 34, 35
technical limitations, 36
Risk communication
　definition, 35, 55
　role of, 3, 56
　stakeholder involvement, 55–56
Risk management, 3
　definition, 22
　process, 22–23
　regulatory system shortcomings, 48
Rivers and Harbors Act of 1899 (RHA), 5, 18, 48, 184

S

Saginaw Bay/River, 92, 120
Screening, particle, 120
Sediment quality criteria, 64
Sediment removal and transport, 82–84
　comparative analysis of remediation technologies, 12–13, 146
　contaminant loss during, 109–112
　cost, 10, 105, 140, 162
　for environmental cleanup, 18–19
　environmental dredging, 104–105
　equipment selection for, 105–109
　as interim control, 90
　for ocean dumping, 25
　on-site controls, 112–113
　recent dredging innovations, 113–116
　storage facilities for, 106
　See also Dredging, environmental; Dredging, navigational; Dredging technology
Sediment traps, 90–91
Separation technologies. *See* Chemical separation; Particle separation
Sewage sludge, 57
Sheboygan River, 127–129, 132
Silt curtains, 112
Site assessment
　acoustic profiling for, 9, 73–74, 77
　aquatic dynamics, 70–71
　chemical sensors for, 9, 74–75, 77

cleanup legislation/oversight, 187–191
contaminant distribution processes, 64–67
core sample evaluations, 72
cost effectiveness, 77
cost of, 9
detailed, 33, 73
for environmental dredging, 106, 107
field surveys, 71–73, 77
goals, 33, 67
identifying decision criteria, 32
for natural recovery, 163
numerical simulations, 75–76
opportunities for improvement, 77, 170
post-project, 82, 139
preliminary, 32
protocol, 67
recent innovations, 73–75
recommendations, 172
remediation costs and, 8–9, 10
sampling design, 72
significance of, for management of sediments, 45, 62, 77, 169–170
use of historical data, 67–70
Soil washing techniques, 118–121, 134
Solids-water separation, 118
Source of contamination
　control challenges, 63, 77
　control strategies, 8
　goals for control, 62–63
　navigational dredging and, 63
　obstacles to identifying, 25
　regulatory control strategies, 63–64, 77
　responsibility, 169
　types of, 15
Stakeholder interests, 7–8
　beneficial uses of sediments, 57–59
　common concerns, 27, 55
　consensus building, 54–55, 60–61, 158, 161
　consideration of, in project planning, 32
　fragmented regulatory system and, 25–26
　off-site mitigation to satisfy, 59–60
　phases of involvement, 53
　range of, 52–53

recommendations for outreach, 158, 161
risk communication among, 55–56
significance of, for management of sediments, 44, 45, 52
threshold issues, 54–55, 259
Sulfide treatment, 99
Superfund, 2, 5, 16–18, 32
  cleanup costs, 27, 63
  evaluation methodology, 6
  in-place capping provisions, 95–97
  natural resource damage claims in, 190–191
  placement decisions, 47
  remedy selection criteria, 188–189
  risk assessment, 156
  sediment disposal regulations, 25
  site inventory, 19, 188
  in situ management, 11
Surfactants, 120
Systems engineering/analysis
  in environmental dredging, 105
  goals, 2–3, 34, 158
  methods, 34
  recommendations for, 158
  risk-based management, 34

T

Tacoma, Port of, 49, 53, 54, 56, 57, 234–236
Temporary interventions. *See* Interim controls
Thermal desorption, 12, 123
Thermal destruction, 123–125, 146
Times Beach, New York, 56–57
Total maximum daily loads, 64, 77, 192–193

Treatment of sediments
  in confined disposal facilities, 132
  cost of, 10
  current state of, 12, 162
  definition, 2 n.2, 16 n.1
  ex situ, 12, 117–118
  in situ, 11
  in situ biological, 100–104
  in situ chemical, 97–100

U

Utility theory, 266–267

V

Vitrification, 100

W

Water Resources Development Act (WRDA)
  of 1986, 7, 19, 50, 56, 59 n.9, 63, 157, 197–201
  of 1988, 201–202
  of 1992, 203–208
Waukegan Harbor, 59, 82, 123, 236–238
Wetlands, 56
WRDA. *See* Water Resources Development Act of 1986

Z

Zeebrugge Harbor, 127
Zero risk, 59